PROTOPLASMATOLOGIA
HANDBUCH
DER PROTOPLASMAFORSCHUNG

HERAUSGEGEBEN VON

L. V. HEILBRUNN UND F. WEBER
PHILADELPHIA   GRAZ

MITHERAUSGEBER

W. H. ARISZ-GRONINGEN · H. BAUER-WILHELMSHAVEN · J. BRACHET-
BRUXELLES · H. G. CALLAN-ST. ANDREWS · R. COLLANDER-HELSINKI ·
K. DAN-TOKYO · E. FAURÉ-FREMIET-PARIS · A. FREY-WYSSLING-ZÜRICH ·
L. GEITLER-WIEN · K. HÖFLER-WIEN · M. H. JACOBS-PHILADELPHIA ·
D. MAZIA-BERKELEY · A. MONROY-PALERMO · J. RUNNSTRÖM-STOCKHOLM ·
W. J. SCHMIDT · GIESSEN · S. STRUGGER · MÜNSTER

BAND VIII

PHYSIOLOGIE DES PROTOPLASMAS

7

AKTIVER STOFFTRANSPORT
(AUFNAHME — TRANSPORT — ABGABE)

a

ACTIVE TRANSPORT THROUGH ANIMAL CELL MEMBRANES

WIEN
SPRINGER-VERLAG
1955

ACTIVE TRANSPORT
THROUGH ANIMAL CELL MEMBRANES

BY

PAUL G. LEFEVRE
WASHINGTON

WITH 31 FIGURES

WIEN
SPRINGER-VERLAG
1955

ISBN-13: 978-3-211-80387-5   e-ISBN-13: 978-3-7091-5768-8
DOI: 10.1007/978-3-7091-5768-8

Protoplasmatologia
VIII. Physiologie des Protoplasmas
    7. Aktiver Stofftransport (Aufnahme — Transport — Abgabe)
      a) Active Transport through Animal Cell Membranes

Active Transport through Animal Cell Membranes

By

PAUL G. LeFEVRE

Medical Branch, Division of Biology and Medicine,
U. S. Atomic Energy Commission, Washington, D.C.

With 31 Figures

Contents

Delimitation of Scope

The province covered by the title assigned to this division of the handbook might reasonably be taken to include an appalling diversity of physiological processes. However, there is no intention here to cover such aspects as the elaboration of special glandular secretions, or the extensive renal and gastrointestinal physiology which might conceivably be included under the heading. A survey of such scope would not only involve a literature of impractically enormous proportions, but would misplace the intended emphasis. The effort here is to cover those lines of investigation which purport to deal more or less directly with the activity of cells in the translocation of substances through the cell surfaces. The transfers concerned will be in general either between the interior and the exterior of the cells, or through layers of cells from one side to the other. Discussion of the cellular extrusion of special secretory products has been avoided; and details of the operation of the special absorptive and excretory organs are taken up only insofar as the experimental approach has been directed toward analysis of the transport phenomena in the various epithelia involved.

In recent years there has been considerable discussion as to what criterion should be used to characterize a transfer system as "active": definitions of varying restrictiveness are presented by Rosenberg (1948), Steinbach (1951 a), Rosenberg and Wilbrandt (1952), Linderholm (1952), and Ussing (1953). Many of the participants in the VIIIth Symposium of the Society for Experimental Biology (1954) also discuss this question (Conway particularly gives here a very convenient tabulation of the possible types of "functional transference"). Since the precise delimitation of the term "active transport" appears to be a semantic rather than a substantive problem, it will not be raised again in this monograph. The attitude adopted is that any translocation process may appropriately be considered here if the role of the cells appears to be more complex than that of merely offering a selective resistance to diffusion.

A few excursions from the strict confines of the title are taken: there is a brief section dealing with the activities of the subcellular particles; also, because of the close bearing on the animal cell studies of certain work with yeasts and bacteria, it is felt essential to include some discussion of the processes in these microörganisms. However, a separate chapter in this handbook will deal with active transport through cell membranes in the plant kingdom generally.

This report will be concerned entirely with actual experimental results in animal systems, and with interpretations developing directly from such experiments, and will thus not include studies of inanimate analogues or general theoretical consideration of the abstract physico-chemical problems raised. Those interested in this aspect of the matter are referred to the excellent treatments of model systems given by SOLLNER et al. (1954), TEORELL (1953), FRANCK and MAYER (1947), DAVSON and DANIELLI (1943), and the interesting suggestions of SPANNER (1954).

HÖBER's classic section on "Passive Penetration and Active Transfer in Animal and Plant Tissues" in his "Physical Chemistry of Cells and Tissues" (1945) and Chapters XI—XIV of DAVSON's (1951) text provide excellent summary discussions of the subject to be covered here. KROGH's (1946) Croonian lecture is also very conveniently organized, but is limited to the inorganic ions. In these publications, the material is arranged in terms of the major anatomical sites of transport functions in the body. In the present review, the presentation will instead be compartmentalized first in regard to the nature of the transported materials, and secondarily in terms of the type and function of the cell or tissue involved. It is hoped in this way to bring out such generalities and specificities as the studies may reveal in comparable activities in different biological systems.

The first and largest section of the discussion will concern transport of inorganic cations, chiefly Na^+ and K^+; later sections will deal with the movement of anions, water, and various organic molecules (mainly the more elementary biological building materials).

Transport of Inorganic Cations

In the interior of most cells, the level of K is many times higher than in the surrounding fluid, and the best evidence indicates that this K is largely, if not entirely, in the form of free ions. Except for cells of the smallest fresh-water organisms, the gradient for sodium ions is generally in the opposite direction, and of the same order of magnitude. Smaller gradients, but perhaps equally high concentration ratios, appear to be usual also for Mg^{++} (concentrated intracellularly), and sometimes for Ca^{++}.

Because of the maintenance of this uneven cation distribution, and the fact that, in media made up of such salts, the animal cells commonly used experimentally show fairly law-abiding osmotic behavoir, it was generally believed twenty years ago that cations could not penetrate the typical animal cell membrane. However, ionic movements into and out of cells were sometimes detected in experimental situations involving upset in the normal ionic pattern of the environment; and chemical studies of these net transfers were beginning to throw doubt on the proposition of general cation-impermeability at the time when isotopic tracer techniques became generally available. The early isotope studies made it clear that the intracellular and extracellular ions are commonly in a state of fairly rapid exchange, and thus made it more difficult to account for the normal maintenance of the high concentration gradients.

A membrane separating two aqueous compartments can of course maintain any pre-existing asymmetry in the distribution of dissolved substances simply by not allowing the ingredients of the solution to pass through from one side to the other. But even though a membrane be easily permeated by ions in the media, its passive properties still can be responsible for an inequality in the ionic distribution. This is seen in such commonly recognized situations as systems in a Donnan equilibrium or with ion-exchange resins; and can be pictured in terms of other simple assumptions as in Ussing's (1947, 1949) "exchange-diffusion" system. In all of these arrangements, there may well be a rapid flux of the ions in both directions, the barrier offering little hindrance to rapid equivalent exchange. Thus the mere persistence of ionic gradients across a biological membrane does not mean that, in interpreting the situation, one is forced to choose between an actively maintained transport or an exceedingly low permeability.

The experimental distinction between these "passive" equilibria and "actively" maintained steady states is seldom easily drawn (there is some doubt that it always even has real meaning). It usually depends on correlation of the ionic distribution behavior with the activity of some metabolic process which can be modified experimentally. The argument may also turn on analysis of kinetics or of potential and flux relations, on demonstration of competitive behavior among related transported substances, or of curious specificities in the system which are not readily explainable on a passive physical basis. Throughout the following presentations, these same basic lines of argument will recur in many special forms, as evidence of the role of cellular active processes in the transfer mechanisms.

K$^+$ and Na$^+$ Transfer in the Red Cell

Metabolism and Uptake of K$^+$.—It was noted many years ago that when mammalian erythrocytes of the K-rich species are stored over a period of days, they rather steadily lose their K$^+$ to the medium. This process had obvious bearing on both scientific and medical interests, and the factors affecting it were given a good deal of attention in the 1930's. It was found that the losses were exaggerated by hypertonicity and by substitution of sugars for the electrolytes of the medium (Ponder and Saslow 1931; Jacobs and Parpart 1933); and that addition of only a small amount of salt markedly reduced the loss in pure sugar solutions (MAIZELS 1935; DAVSON 1939). WILBRANDT (1940 b) pointed out that generally these adjustments to alterations in the medium showed a rapid, *reversible* phase followed by a slower, progressive, and irreversible loss of the cell contents (and hence decreased osmotic "fragility").

It also seems to have been WILBRANDT (1940 a) who provided the first evidence that the K+ retention is associated with the intactness of a definite metabolic system in the cell. Others had noted effects with certain metabolic inhibitors: ØRSKOV (1935) had shown that Pb-poisoning[1] of some species of erythrocytes (not others) could lead to a loss of K+ in exchange for Rb+; and DAVSON and DANIELLI (1938) had induced prelytic K+ loss with various alcohols and other agents, but had found no such effect with more specific inhibitors such as NaF, NaCN, or CO. WILBRANDT found, however, that the loss of K+ from human red cells at 37° C. in the presence of NaF at 0.04 M was so severe that after a few hours the cells failed to hemolyse even in only 1/10-isotonic NaCl[2]. This represented a K+ loss at about 25 times the rate observed in the absence of the poison. The glycolysis-inhibiting property of fluoride, rather than its Ca++-removing property, was implicated by the fact that the action was shared by iodoacetate but not by oxalate or citrate. This interpretation was further supported by the facts that (1) paralleling the action of fluoride on K+ retention, there was an inhibition of the rate of glycolysis; (2) all the tested species of strongly glycolytic K+-rich red cells were affected, whereas weakly glycolytic species were not; and (3) pyruvate prevented the fluoride effect on K+ loss[3].

A highly significant extension of WILBRANDT's findings was soon made by HARRIS (1941) and DANOWSKI (1941). They independently observed that, following loss of K+ from human red cells stored at refrigerator temperatures, there was partial recovery of the lost K+ (against the concentration gradient) when the cells were returned to body temperature; and glucose assisted in this recovery. Table I shows some representative data. The K+ uptake persisted for only a short time after the cessation of glycolysis as measured by the disappearance of serum glucose and the appearance of inorganic phosphate. Thus the glycolytic process evidently not only served somehow to maintain the K+ distribution, but also tended to restore it after it had partly run downhill. By analysing the dynamics of the net K+ loss in terms of the simplest assumptions as to the kinetics, PONDER (1949, 1950)

---

[1] Pb++ itself may be actively taken into the red cell; at any rate, MORTENSEN and KELLOGG (1944) found that Pb++ (as Radium D) becomes rapidly and reversibly attached to red cells up to forty times the level in the medium, the process showing an apparent activation energy on the order of 21,000 cal./mol.

[2] Later such cells undergo colloid-osmotic hemolysis as Na+ entry begins.

[3] Other details of the observations indicated that the response of the K+-retaining system is not simply a direct consequence of the depression of glycolysis. An analysis in DAVSON's (1941) confirming report suggests that the fluoride effect is attributable to accumulation of an intermediary metabolite which alters the cell membrane's properties.

made an estimate of the apparent rate of the K^+-accumulating process. At all temperatures considered (4–37° C.), he calculated that the observed glucose utilization was far more than enough to supply the energy needs for the K^+ uptake.

Table I. *Glycolysis and K^+ Loss from Human Red Cells.*
(Adapted from Danowski, 1941.)

Hours of incubation at 37° C.	Serum K+, in mEq/L			
	Control	Glucose added at 2.5 hrs.	Control	NaF, 0.05% added at 0 hrs.
0	4.1	4.1	4.0	4.0
2.5	3.4	3.4	3.5	6.1
5	3.1			
8.5	4.0	2.9		
16.5	8.0	3.3		

In the chicken erythrocyte, Maizels (1954) reports that the maintenance of cation distribution is not particularly dependent on glycolysis; it is instead highly sensitive to cyanide, azide, or DNP [4], so that an association with aerobic respiratory processes is implied. In incomplete preliminary experiments, similar behavior was seen in the snake and tortoise, so that it is probably characteristic of the nucleated erythrocytes.

The Relative Roles of Na^+ Extrusion and K^+ Accumulation.—If Na-rich red cells (such as those of dog or cat) are suspended in high-K^+ solutions, a slow exchange of Na^+ and K^+ occurs, similar to that seen in stored human blood, but in the opposite direction. Studying cat cells in various mixtures of KCl and NaCl, Davson (1940) found a number of differences between the K^+ entry and the Na^+ exit processes, as regards their sensitivity to the composition of the medium. Amplifying on these findings, Davson and Reiner (1942) argued that the response to acidity, temperature, tonicity, narcotics, and heavy metals points in this cell to the Na^+ transfer system, rather than the K^+ transfer system, as metabolically sustained. This viewpoint was for a time championed by Maizels (1948, 1949) as applying even to the K^+-rich human red cell; Maizels pointed out that it might be a mistake to interpret the observed uptake of K^+ as a primary active process, since it could result passively from an active expulsion of Na^+. He believed such an active extrusion was demonstrated by the failure of the heparinized cells to swell in a Na_2CO_3 medium.

Flynn and Maizels (1950) made use of Li^+ as a replacement for Na^+ in the medium to elucidate further the supposed primary nature of the Na^+-extrusion process in the human red cell. Uptake of K^+ failed to occur

---

[4] 2, 4-dinitrophenol.

when Li+ was available to replace the extruded Na+. The Li+ taken in was apparently not then actively extruded. The backflow of Na+ into the cells decreased with its level in the medium, and when this was less than about 8 times the intracellular level, the expulsion exceeded the backflow so that an actual net loss of cellular Na+ occurred. However, these findings were not confirmed by PONDER (1950), who reported that the uptake of K+ was depressed only by 25–50% when 80% of the medium's Na+ was replaced by Li+ or Cs+. PONDER felt that the extrusion of Na+ could not be the critical process in determining the K+ uptake.

HARRIS and MAIZELS (1952) abandoned the notion of the primacy of Na+

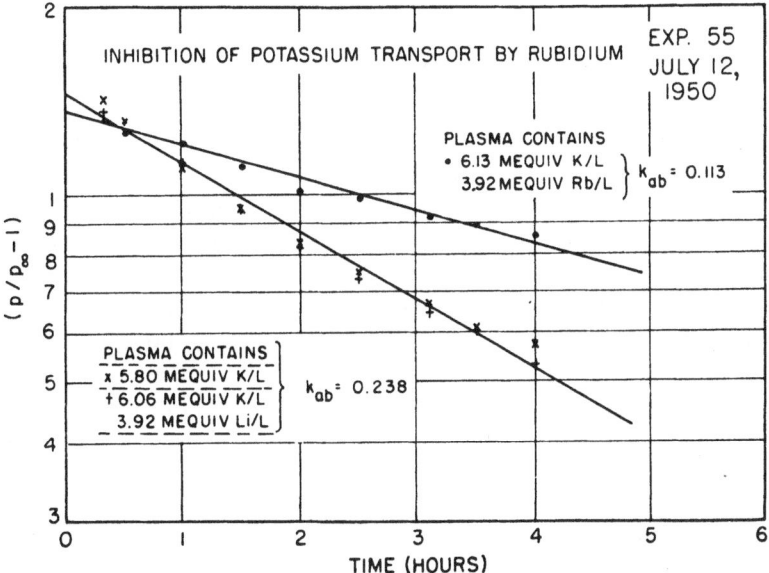

Fig. 1. Uptake of K⁴² from plasma by human red cells as affected by Rb+ and Li+. Tracer K⁴²Cl added at time "zero". p = concentration of K⁴² in the plasma; p∞ = p after equalization of inside and outside specific activities. Theory predicts that ln(p/p∞ −1) will fall linearly with time, the slope and y-intercept permitting calculation of Kₐb, the transfer coefficient in the direction, plasma → cells.

(Courtesy of A. K. SOLOMON and Journal of General Physiology)

extrusion on the basis of the fact that the distribution of K+ did not correspond to the apparent electrochemical potential across the red cell membrane. This potential was taken to be defined by the Cl−, H+, or HCO_3^- distribution; $[Cl^-]_{ext.}/[Cl^-]_{int.}$ and $[H^+]_{int.}/[H^+]_{ext.}$ varied together from a normal value of about 1.4, as the medium was mildly altered, but did not respond to the cell's metabolic state. In contrast, $[K^+]_{int.}/[K^+]_{ext.}$ is ordinarily about 35, so that the intracellular excess is far too great to be attributed to the electrostatic forces which are presumably partly the result of the transport of Na+. Moreover, SOLOMON (1952) found the apparent activation energies for movement of both cations in each direction were very much higher than for diffusion in aqueous solution [5]. SOLOMON brought

---

[5] CLARKSON and MAIZELS (1954) find apparent activation energies for the Na+ system at variance with SOLOMON's figures, but not altering this argument.

out several differences in apparent activation energies and in concentration-dependency, in comparing the movements of Na+ and K+, and pointed out that there was no apparent cross-competition between the two systems. He showed that Rb+ and K+ compete for one transport pathway, while Li+ and Na+ compete for another (with approximately equivalent potencies). Fig. 1 illustrates the depressed rate of intracellular migration of K^{42} in the presence of Rb+, and the insensitivity of the system to addition of Li+. Tosteson and Dunham (1954) have also studied the action of Cs+ in the K+-transport system. Although Cs^{137} entry into the human red cell was depressed at higher K+ levels, the ratio of the rate constants for the two ions varied with their relative levels in the medium. Thus there is both a competitive and a noncompetitive component in the total fluxes [6].

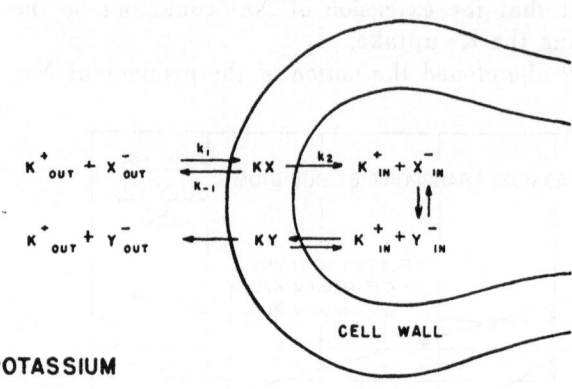

POTASSIUM

WORKING HYPOTHESIS FOR CATION TRANSPORT
IN HUMAN RED BLOOD CELLS

SODIUM

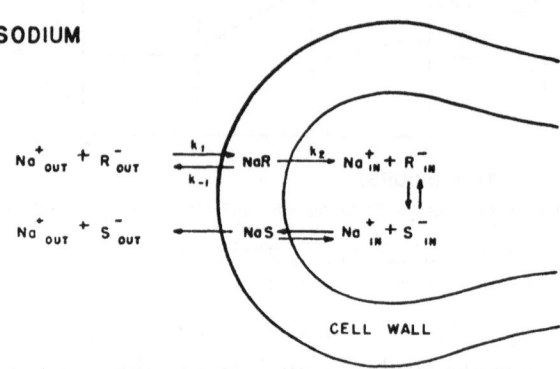

Fig. 2. (Courtesy of A. K. Solomon and Journal of General Physiology)

Solomon's observations led him to suggest a two-way carrier system for each of the two ion groups, as in the diagram of Fig. 2, such that the inward and outward transports are linked by the intracellular conversion of the inward carrier to the outward carrier. This metabolic step could thus indirectly govern the operation of the overall transfer. The chief difference in the two hypothetical "ferries," demanded by the experimental behavior, is that the Na-ferry must be not nearly saturated in the normal environment, while the K-ferry is loaded to near capacity. Harris and Maizels (1952) suggested even that the two ferries are bound in a series reaction, the carrier undergoing a cyclic transformation between a K-carrying form and a Na-carrying form, passing alternately in and out of the cell without being released; the evidence for this notion is discussed by Harris (1954).

---

[6] This report suggests incidentally that the process of sickling in human erythrocytes entails an acceleration of the active cation-transport process.

According to Maizels's (1949) estimations of "cell pH" in hemolysates, the Na$^+$-extrusion system has a rather distinct optimum at pH 7.3–7.4. The correlation with glycolysis is stressed in a later report (Maizels 1951), which showed that the Na$^+$ expulsion does indeed involve the liberation of acid metabolites and a drop in phosphoric esters in the cell, as illustrated

Table II. *Relation of Glycolysis to Cation Movements in Incubated Red Cells. Effects of Inhibitors and Substrates.*
(Compiled from data of Maizels, 1951.)

Additions to medium		Change in cell contents during incubation				
NaOH	Inhibitor	K	Na	Acid-sol. P	Easily hydr. P	Glucose [1]
mM	*mM*	*mM*	*mM*	*mg. %*	*mg. %*	*mg. %*
4.2	None . . . . . —	29	— 40	— 13	— 1.8	— 192
—	NaF 10	— 18	24	— 6	— 3.6	0
0.6	NaIAc 0.8	— 6	20	— 29	— 5.2	— 10
—	NaCN 10	11	— 15			
5	NaN$_3$ 8	18	— 23	— 11	— 0.7	— 192
4.5	Mepacrine . . . 0.5	13	— 9	— 17	— 0.1	— 172
4	Na$_2$HAsO$_3$ . . . 2.8	14	— 17	— 12	— 0.7	— 142
7.3	DNP 2.4	20	— 23	— 16	— 1.9	— 238
4.5	Na malonate . . 10	27	— 34			
6.7	Methylene blue, 12 mg./L.	26	— 32	— 5	— 0.7	— 256

Substrate replacing Glucose						Cell sugar [1]
						mg. %
None		0	7	— 26	— 2.7	—
Galactose		2	2	— 25	— 2.9	— 18
Fructose		29	— 30	— 9	2.7	— 136
Mannose		26	— 31	— 11	— 0.5	— 142
Pyruvate, 20 mM		0	5	— 24	— 2.4	—
Arabinose		3	— 5	— 28	— 3.6	— 12
Xylose		2	— 2	— 25	— 4.1	— 6
Lactose		5	— 2	— 21	— 2.9	— 10
Maltose		8	— 5	— 23	— 3.0	— 24
Sucrose		— 1	6	— 22	— 2.7	—

[1] Control cell sugar determinations were made with addition of 30 mM NaF, to stop all glycolysis. Incubations were for 18 hours, with initial extracellular K$^+$ at 33–35 mM, Na$^+$ at 115–120 mM. Monosaccharide levels used were on the order of 300 mg. %; disaccharide levels, 110 mg. %.

in Table II. Note that mannose was as effective as glucose in maintaining the system, and fructose at very high concentrations was also usable; but galactose and a variety of other likely substrates were ineffectual. Fluoride or iodoacetate blocked Na$^+$ expulsion, but cyanide, azide, CO, DNP, and other special inhibitors did not.

Harris and Maizels (1951), calculating conventional velocity constants for the Na+ movements on the basis of the assumption that the uni-directional fluxes were essentially monomolecular, found the outward trans-fer constant to be about 10 times as large as the inward transfer constant at 37° C. As would be anticipated from the general picture developed above, the outward flux showed far greater temperature-sensitivity, and at 4° C. both constants were on the order of 0.004 per hour; thus at this temperature there would be a net entry of Na+ until the intracellular and extracellular levels were equal. In general, both velocity constants were qualitatively affected similarly by experimental variables, but fluoride-poisoning depressed the extrusion without reducing the rate of entry[7].

Exchangeability of Intracellular K.—The earliest radioisotopic studies of cation movements through the red cell surface (Hahn, Hevesy, and Rebbe 1939; Eisenman *et al.* 1940) showed such a slow exchange for K+ that it was concluded that it was not in dynamic equilibrium at all; Na+ exchange was only slightly faster. However, these observations were soon generally rejected on technical grounds; an exchange rate of about 1.5% of the cell K per hour, such as reported by Dean *et al.* (1941) has been the experience of subsequent investigators. Hald *et al.* (1947, 1948) were led to revive the doubt that Na+ or K+ can actually *permeate* the membrane of a resting red cell, since their chemical analyses on human blood which had been incubated with added solutes, with unusual precautions to maintain as nearly undisturbed conditions as possible, showed virtually *no* ionic ex-changes, in the face of unusual gradients. Since such a high degree of physical impermeability was apparent, they concluded that any such movements observed in undamaged cells involved some sort of special translocation process.

In Ponder's (1949, 1950) analysis of the dynamics of the K+ loss, mentioned earlier, stress was laid on the fact that the level of intracellular K, toward the end of the exponential loss, approached an asymptote at a level considerably higher than the extracellular [K+]. The glycolytic in-hibitors lowered the asymptote-level, but still left an apparent substantial accumulation even in F⁻-poisoned cells at 4° C. in the absence of glucose. Ponder thus concluded this was not the result of an uptake process, but that some of the intracellular K was in an "immobile" form. Later Ponder (1951) found that even the mobile K-fraction was apparently not homo-geneous, but showed two phases of exponential loss, of very different rates.

But the characteristics of K behavior as analysed by its net loss over these relatively long periods are not duplicated in experiments with K⁴² tracing techniques, which permit more nearly normal conditions. Raker *et al.* (1950) found with this method not only a rapid turnover and attain-

---

[7] Mention should be made of the fact that this group (Blowers, Clarkson. and Maizels 1951) has found that the action of a great number of variables on the maintenance of the Na+-extrusion system correlates in detail with their effect on the persistence of the unexplained "flicker phenomenon" in the red cell.

ment of essentially complete equivalence of specific activities of intracellular and extracellular K within 36 hours, but also that the time-course
of the specific activity changes fit the suggestion of a single extracellular
and a single intracellular compartment of K (a linear drop in the logarithm
of the difference between the two specific activities). Increase of the extracellular [K+] up to 74 mM or variation of pH between 7.0 and 7.7 had no
appreciable effect on the transfer rate. SHEPPARD and MARTIN (1950) confirmed that this process followed simple two-compartment dynamics.
However, a report soon to appear from SOLOMON and GOLD (1953) indicates
that, although such a system is adequate to describe the *influx* of K^{42}, the
efflux characteristics imply a three-compartment system, with about 5%
of the intracellular K in a more rapidly exchanging compartment. It is at
present difficult to say to what extent the heterogeneity of the red cell
populations dealt with experimentally might account for this complication.

SHEPPARD, MARTIN, and BEYL (1951), comparing the red cells of sheep,
dog, cow, and man, found by similar methods that the K+-exchange rate
correlates roughly with the intracellular content of the ion. The Na+ exchange is more rapid, relative to the intracellular levels; the Na in human
cells appeared to be in at least two compartments of roughly equal size,
differing in their exchange rates. SOLOMON's (1952) report provides a
systematic confirmation and definition of the major characteristics of the
cationic movements in the human red cell as revealed by the use of Na^{24}
and K^{42}.

Possible Role of Cholinesterase Activity.—An extensive series of papers
from Vanderbilt (GREIG and HOLLAND 1949 a, b, 1951; HOLLAND and GREIG,
1950 a, b, 1951; LINDVIG, GREIG, and PETERSON 1951) have reported an interesting correlation between the activity of the acetyl cholinesterase system
and the movements of Na+ and K+ through the erythrocyte membrane,
particularly of man and dog. These studies involve following hemolysis
(in some cases also prelytic swelling or movements of K+ as in Fig. 3) in
various mixtures of Na- and K-salts, as a function of the addition to the
system of acetyl choline, eserine, and a variety of other drugs affecting
cholinesterase activity. The conclusiveness of these experiments is in
general rather disappointing. In most instances, the make-up of the
suspending medium was markedly different from that of normal plasma
(as in the example of Fig. 3), and the effects observed show numerous
unexplained peculiarities that greatly complicate the details of interpretation. Nevertheless, there emerges the definite general pattern: pro-
vision of a substrate for the cholinesterase system tends to preserve the
cells against upsets in the normal electrolyte pattern (and the resultant
hemolysis) which ensue in the absence of such a substrate, or upon addition
of an agent preventing the cholinesterase from acting.

GREIG, MAYBERRY *et al.* (1951, 1953) reported that acetyl choline or its
relatives could even induce the replacement of K+ after its loss from
human red cells during incubation (this recovery did not however include
extrusion of extra Na+ that had entered the cells). Physostigmine blocked

this process, as well as the K+ recovery initiated by addition of glucose. However, Christensen and Riggs (1951) raised serious question that this involves the inhibition of cholinesterase. They observed that cationic

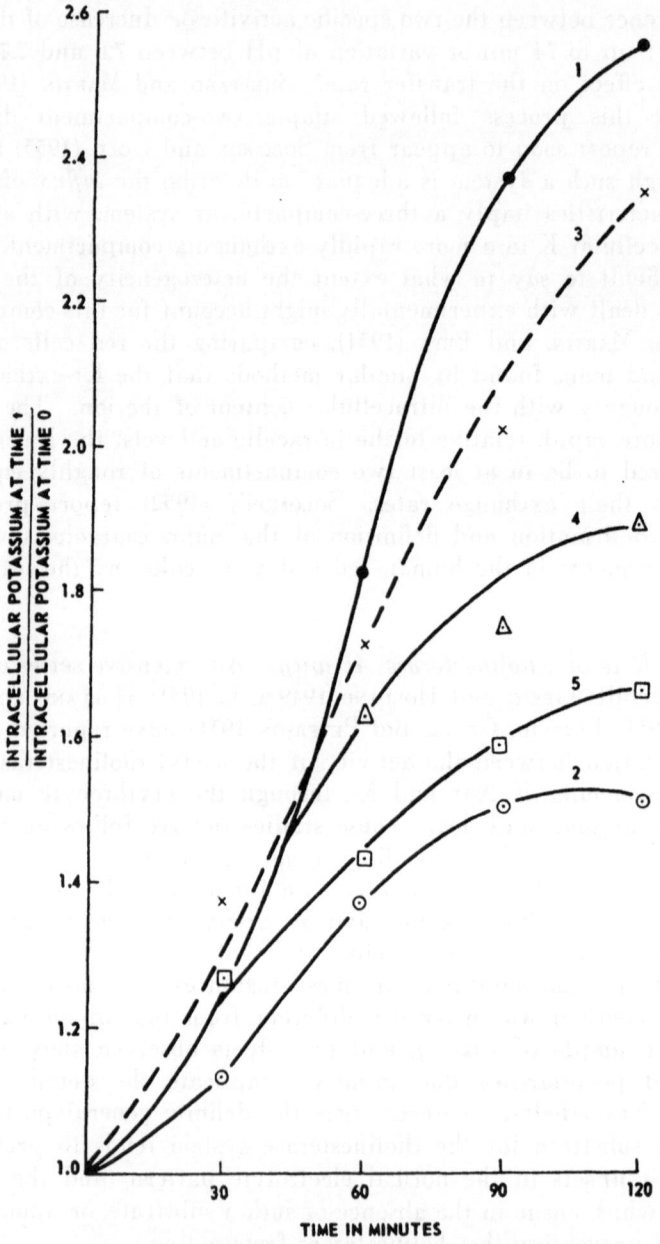

Fig. 3. Uptake of K+ by dog erythrocytes as a function of cholinesterase activity. Cells suspended at 5° C. in isotonic mixture, approximately 2 KCl : 1 KHCO₃. 1. Without acetyl choline
2. Acetyl choline, $5 \cdot 10^{-3}$ M
3. A. Ch. plus Physostigmine, $2.7 \cdot 10^{-5}$ M
4. ,,   ,, $5 \cdot 10^{-6}$ M
5. ,,   ,, $1 \cdot 10^{-6}$ M

(Courtesy of Archives of Biochemistry; from Holland and Greig, 1950 a) physostigmine actually accumulates in duck or human red cells and in the free cells of the mouse Ehrlich ascites carcinoma, replacing (or displacing) K^+. However, this accumulation does not appear to involve an active transport, but rather a fixation of the drug as an indiffusible cation.

PARPART and HOFFMAN (1952) have also challenged the interpretation of GREIG's work, claiming that the effects attributed to acetyl choline can be duplicated by acetic acid, such as would appear upon hydrolysis of the added acetyl choline; but GREIG claims that the dissimilarities in the respective experimental designs make this criticism invalid. In any case, it would appear premature at the present stage to assign to the cholinesterase system a function in the cation distribution system, but this line of investigation shows promise of interesting development. TAYLOR and WELLER (1950), and TAYLOR, WELLER, and HASTINGS (1952), examining the system in human blood under more normal conditions than in the experiments discussed above, found with K^{42} that, at the lowest effective concentrations, cholinesterase inhibitors decreased K^+ influx, whereas cholinacetylase inhibitors increased K^+ efflux. But the concentrations required were somewhat higher than those necessary to inhibit the enzymes.

Na⁺ and K⁺ Transport in Muscle

Turnover of Potassium.—Classical electrophysiological theory interprets the resting potential across the plasma membrane of nerve and muscle cells on the basis of an appreciable permeability of the membrane to K^+, with a virtual impermeability to Na^+ and to the bulk of the intracellular anions. In keeping with this idea, early radioisotopic studies with K^{42} (HAHN, HEVESY, and REBBE 1939) showed in the intact rabbit or frog an appreciable, though fairly slow, exchange of tissue and plasma K^+, not only in the muscles but in many other tissues. FENN, NOONAN, MULLINS, and HAEGE (1942) found that muscle potassium was completely exchangeable: this has been well confirmed, especially in the rat diaphragm preparation (CREESE 1951; CALKINS *et al.* 1954)[8].

The order of magnitude of the concentration ratio maintained is physicochemically appropriate to the size of the resting potential, so that on this basis there is no reason to question that the K^+ distribution represents a simple physical equilibrium. Moreover, HARRIS and BURN's (1949) finding that the K^+ "permeability constant" appears not to depend on the extracellular $[K^+]$ is in keeping with this. Similarly, CONWAY, CAREY, and MOORE (1950) found that the K^+-entry half-time in frog sartorii in mixtures of KCl and NaCl did not depend on the proportions used, which means that the absolute rate of K^+ entry is proportional to its concentration in the medium. When the Cl^- was partly replaced by a larger anion, it was evident that the rate depended on the product, $[K^+] \times [Cl^-]$. Moreover,

---

[8] However, WESSON *et al.* (1949) found two compartments in chicken embryo muscle K, differing in rate and in temperature-sensitivity. HARRIS (1953 a) claims that frog sartorii also show at 18° C. a small non-exchangeable K-component which becomes more prominent at lower temperatures.

Dean (1940) found that, except for a few dying fibers, anoxia did not alter the maintenance of cell K in excised frog muscles, although constant exchange with the K in the medium was shown. Similarly Fenn, Koenemann, and Sheridan (1940) could not show any evidence of loss of K^+ into an anoxic perfusate which was run through the vessels of frogs' legs. All these facts would suggest no metabolically limited step in the maintenance of the resting K-distribution.

In the rat diaphragm, Calkins et al. (1954) also noted with K^{42} a monomolecular type of K^+ loss. They found that the rapid K^+ exchange rate of about 1.5% of the contents per minute did not involve any net loss in the presence of glucose. While lowered temperature decreased the inflow, the fact that anoxia lowered the gradient at first only by increasing the outflow suggested that the intracellular level was not set primarily by action of any inward "pump" mechanism. Creese (1952) noted that K^+ loss from this same preparation was brought on by removal of bicarbonate, even without disturbance of the pH, and that restoration of the HCO_3^- led to recovery of the lost K^+. Verzár and Somogyi (1939) found that the disappearance from cat blood of injected rapidly absorbed sugars was accompanied by a rise in blood $[K^+]$, although a net *uptake* of K^+ accompanies glycogen deposition in the rat diaphragm (Calkins et al. 1954).

Though these latter observations are suggestive, none seems to *require* an active K^+ transport. However, Ling (1953) showed that uptake of Rb^+ into frog muscle at various $[K^+]$ followed the Lineweaver-Burk theoretical pattern for competitive inhibition in a system of the Michaelis-Menten form, so that these two ions apparently share the same transport pathway, as they do in the red cell. Moreover, Keynes (1954) finds that the total conductance of frog muscles is very much higher than the partial K^+ conductance calculated from the fluxes determined with K^{42}; this would certainly seem to show that independent diffusion of the individual ions is not the basis of these fluxes.

Extrusion of Na^+.—Although (except perhaps for these latest observations) the observed movements of K^+ do not necessarily belie classical theory, there is strong evidence against the postulated Na^+-impermeability. Heppel (1939, 1940) showed that the muscles of rats on a diet poor in K gradually replaced up to half their K with Na, and that this situation was rapidly restored to normal when K was restored to the diet (see also Conway and Hingerty 1948). Similar exchanges *in vitro* between muscle cells and the medium were observed by Steinbach (1940). Thus an appreciable Na^+ permeation was suggested as the normal situation, the low intracellular level being maintained by some sort of extrusion mechanism.

Conway (1946) has attacked this idea vehemently; in his experiments, the osmotic behavior of frog sartorii in Ringer's solution in which small amounts of the Na^+ were replaced by increments of K^+ was such that a free permeability to K^+ (with a half-time of about 40 minutes at room temperature) was indicated. If the permeability to Na^+ were of this order of magnitude, the entire resting energy metabolism of the muscle would not suffice to maintain the observed degree of Na+ exclusion. Conway agreed that muscles did acquire Na+ during long exposure to K+-free NaCl solutions at 3° C.; but he did not believe this reflected depressed activity of an extrusion process, for analysis of these muscles after treatment with KCl-NaCl mixtures showed that the Na+ which had gained entry was not extruded in connection with the K+ reëntry, and their osmotic behavior was altered accordingly.

Steinbach (1951 b) extended his earlier observations on sartorii recover-

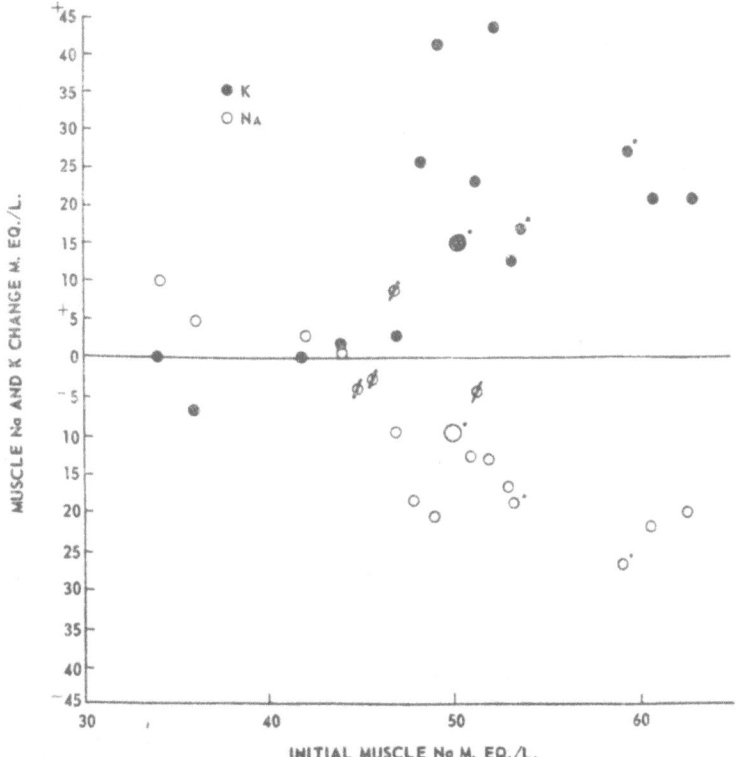

Fig. 4. Recovery of K+ and extrusion of Na+ in frog muscle as a function of initial Na content. After 12–18 hrs. in low-K+ Ringer's at 5°C., one of each pair of sartorii was analysed for initial level, and other was soaked at 20°C. for 5–7 hrs. and analysed for ionic recovery. Solid dots are K concentration changes; open circles are Na concentration changes, the split circles being Conway's data, plain circles being Steinbach's data. Large symbols are averages from earlier work (denoted by accents) and Conway's data; small symbols are on individual muscle pairs.

(Courtesy of H. B. Steinbach and American Journal of Physiology)

ing from long exposures in the cold to low-K+ media, and showed that the discrepancies between his and Conway's findings regarding the extrusion of Na+ disappear if all the data are plotted as in Fig. 4, as a function of the intracellular Na+ level at the end of the leaching-out period (at the beginning of the experimental period). As the figure shows, if this level is less than about 45 mM, there is little if any recovery following restoration to room temperature and addition of K+ to the medium; whereas, the higher above this level the intracellular [Na+], the greater the subsequent Na+ extrusion and K+ reaccumulation. By substituting choline chloride for part of the NaCl in the K^+-depleting solution, STEINBACH (1952) showed that the extrusion of Na^+ was the active process; no significant recovery of K^+ occurred in muscles whose K^+ had been depleted without corresponding Na^+ entry. That the choline itself was not responsible for this was indicated by the fact that Na^+-loaded muscles had no difficulty in unloading the Na^+ into a choline$^+$-K^+ medium [9].

Table III. *Relation of Bioelectric Behavior to Muscle Ionic Distribution Maintained by Sodium Extrusion.*

(Adapted from DESMEDT, 1953.)

Treatment of Muscle	Resultant Intrafibral Level of		Membrane Potential			
	Na+	K+	Resting Level	Peak of Activity (Overshoot)	Max. rate of Rise	Max. rate of Fall
	mM	*mM*	*mV*	*mV*	*V/sec.*	*V/sec.*
Soaked 31 hrs. in low-K+ Ringer's	55	111	82 ± 1	16 ± 1	372 ± 20	66 ± 3
Same, then soaked 90 min. in Ringer's with K+ at 10 mM	32	139	93 ± 1	29 ± 1	394 ± 16	98 ± 4

All tests performed in phosphate-Ringer's with K^+ at 2.5 mM, at 20^0C. Voltages are means ± standard errors, with 7–8 determinations in each case.

The *net extrusion* measured by STEINBACH was not very different from the *total exchange* (unidirectional efflux) determined with Na^{24} by USSING (1947) and LEVI and USSING (1948). Allowing for a very rapid (presumably extracellular) phase, they found the tracer washed out of the sartorii with a half-time of about 35 minutes, without regard to the outside $[K^+]$ [10]; the Cl^{38} exchange was even more rapid. By CONWAY's system of calculation, the Na^+-efflux energy requirements come to about half of the total resting metabolic rate and make it dubious that this is entirely a metabolically sustained transport [11]. LEVI and USSING invoke an "exchange-diffusion"

---

[9] DESMEDT (1953) has verified STEINBACH's observations of a net extrusion of Na^+ against a gradient when K^+ is added at about 10 mM. DESMEDT further noted (Table III) by use of Ling microelectrodes that the action potential "overshoot" in these sartorii was set by the existing Na^+ gradient, in agreement with the theory developed by the Cambridge school in connection with neurone function (discussed in a later section).

[10] KEYNES and MAISEL (1954) note that the extrusion is faster with $[K^+]$ at 10 mM than in a K^+-free medium; KEYNES (1954) found that a preliminary K^+-free wash is necessary to bring out the K^+-dependence dramatically. Thus an extremely low level of K^+ is evidently required to show any effect of $[K^+]$ on the Na^+-extrusion system.

[11] KEYNES and MAISEL (1954) however found by using paired sartorii for the two types of determination that the extrusion would take only 10% of the resting metabolic energy; they attribute LEVI and USSING's higher figure to the comparison of dissimilar muscle preparations.

system as a reasonable substitute; this involves essentially simply a reversible combination of the Na^+ at an attachment site on a molecule restricted to the membrane, but free to diffuse within it from one side to the other.

HARRIS and BURN (1949) carried out a thorough mathematical analysis of the Na^+ and K^+ transfers between the medium and a thin sheet of muscle cells, and by the use of radioisotopes estimated the rates in frog sartorii in terms of conventional "permeability constants"; these proved to be on the order of several microns per hour. The constant for K^+ was about seven times that for Na^+. HARRIS (1953 b) negated the suggestion that the extrusion might be energized by high-energy P-bonds in such a way as to involve loss of free phosphate, since the ratio of Na/P expelled by frog muscle into an initially phosphate-free medium was over 3 times the maximum possible in terms of the work requirements on this basis. Moreover, KEYNES and MAISEL (1954) found that DNP, cyanide, and iodoacetate are without serious effect in this system.

Possible Role of Cholinesterase Activity.—HOLLAND, DUNN, and GREIG (1952 a, b) have investigated the applicability to heart muscle of their ideas regarding the role of the cholinesterase system in cation movements, as discussed in connection with erythrocytes. They found that the presence of acetyl choline or other suitable substrate enhanced the loss of K^+ from young guinea-pig auricles in K^+-free Tyrode's solution, and the concentration-dependency closely paralleled that of the concurrent hydrolysis of the acetyl choline. Later recovery of K^+ from a balanced solution was also somewhat accelerated by acetyl choline, and physostigmine prevented these effects.

The condition of myasthenia gravis, in which the weakness is temporarily relieved by prostigmine, is interesting in this connection. CUMINGS (1940) found that prostigmine, although without effect on normal human K^+ distribution, promotes in myasthenics a temporary loss of muscle K^+ into the plasma. But THOMPSON and TICE (1941) found this action to be highly variable, depending on the extent of pretreatment of the patient with K salts. They also showed that prostigmine injection significantly increased the muscle K^+ content in rats.

Shifts with Muscular Activity.—FENN and his associates (1936, 1937) observed that rat and cat muscles, stimulated tetanically through the nerve for several minutes, lose appreciable amounts of K^+ in exchange for Na^+, and that the normal picture is restored (against the gradients) during the next few hours [12]. In addition to the presumably intracellular Na^+ which replaced the K^+, there was an uptake of NaCl and water in proportions corresponding to a slight hypotonicity, which presumably added simply to the extracellular spaces. FENN (1937) also showed such behavior follow-

---

[12] Frog muscles behaved similarly upon *direct* stimulation. but the reaction under neural stimulation was debatable.

ing voluntary contraction in swimming rats, comparing the muscles of intact and denervated legs in the same animals [13]. Wood, Collins, and Moe (1940) made use of a dog heart-lung-gastrocnemius preparation to study these phenomena with shorter bursts of activity of a more normal character. Under such circumstances, analysis of the blood content changes showed the recovery transfer processes were rather immediate upon cessation of the activity, and proceeded at a rapid rate except for the change in hydration. Here then is an apparently noteworthy operation of the muscle cation transport system in a common functional situation. But no further analysis of the mechanics of the recovery processes has been reported. The basic phenomenon has been observed in human patients, also, in connection with electroshock therapy (Welt et al. 1950); with convulsions, there was a transient rise in blood $[K^+]$ and a lesser rise in $[Na^+]$, $[Cl^-]$, and [protein], so that there was evidently a loss of K^+ from muscle and a transfer of water into the cells. In the recovery phase, K^+ apparently reënters the cells *beyond* the initial resting level.

Cicardo and Moglia (1940) observed that acetylcholine-induced contracture in toad leg muscles was associated with a marked loss of K^+ into the perfusate, especially if the muscle were denervated; curare blocked this action. Similarly, Verzár and Somogyi (1940) found that injection of acetyl choline caused a considerable elevation of the $[K^+]$ in the venous blood from the leg of a cat, like that seen when the sciatic nerve was stimulated.

Na$^+$ and K$^+$ Transfer in Nerve Cells

Metabolism and the Maintenance of Resting Ionic Distribution.—The basic situation in regard to the maintenance of the K^+ and Na^+ gradients, and their relation to the resting bioelectric potential, appears to be essentially similar in nerve to that discussed above for muscle; but the experimental study of the ionic movements had not been nearly so extensive until the last few years. Like most other animal cells, nerve cells lose appreciable amounts of intracellular K^+ to the medium under anaerobic conditions (Fenn and Gershman 1950), and glucose can prevent or even partly reverse this loss, as Dixon (1949) showed in slices of rabbit cerebral cortex. According to van Haareveld (1950), rabbit sciatic nerves to which glucose and oxygen are available can maintain their internal $[K^+]$ for at least 6 hours at 37° C. in Ringer's. He noted that the application of arsenite at 10^{-3} M was comparable to anoxia in allowing loss of K^+. Shanes (1951) made the interesting observation that the loss of excitability in a nitrogen atmosphere, in squid giant axons or in crab nerves, could readily be temporarily relieved by washing with O_2-free sea water, as if the source of the block were the exuding from the cell interior of a depressing agent, very likely K^+.

Terner, Eggleston, and Krebs (1950) found that effective maintenance

---

[13] With the prolonged quiescence of denervation, the muscles showed a slight gain in K^+.

of brain slice K^+ required not only O_2 and glucose, but also either glutamate or aspartate; this group reported also (KREBS *et al.* 1951; DAVIES and KREBS 1952) that cold storage or O_2-deprivation led to a loss of K^+ and gain of Na^+ in ox retinal fragments, this exchange being prevented by glucose and reversed upon incubation with glucose and glutamate. Table IV

Table IV. *Effects of Substrate and Inhibitors on Retention of K+ in Neural Tissues.* (Adapted from TERNER, EGGLESTON, and KREBS, 1950.)

Conditions and Additions During Incubation (Concs. in mM)	Tissue [K+], in mEq./kg.		
	Initial	Final	Change
Rabbit brain-cortex slices	98.6		
Aerobic; no addition		28.6	— 70.0
glucose, 20		38.8	— 59.8
glucose, 20; L-glutamate, 10		68.2	— 30.4
Anaerobic; no addn.		10.4	— 88.2
glucose, 20		38.3	— 60.3
glucose, 20; L-glutamate, 10		22.7	— 75.9
Ox retinal fragments; aerobic	32.8		
No addition		29.9	— 2.9
Glucose, 20		35.1	2.3
Glucose, 20; L-aspartate, 10		43.8	11.0
Glucose, 20; L-asparagine, 10		32.7	— 0.1
Glucose, 20; L-glutamate, 10		55.0	22.2
Glucose, 20; L-glutamine, 10		31.7	— 1.1
Ox retinal fragments; aerobic	34.3		
Glucose, 20		42.4	8.1
Glucose, 20; L-glutamate, 10		68.3	34.0
Both glucose and glutamate			
+ fluoride, 20		35.8	1.5
+ malonate, 20		55.0	20.7
+ iodoacetate, 1		8.6	— 25.7
+ DNP, 0.2		41.0	6.7

All incubations in bicarbonate-saline for one hour.

presents some typical observations. Stimulation of ionic recovery was perceptible with glutamate at only 5.10^{-4} M, and the rate of K^+ uptake paralleled the glutamate concentration. The uptake of glutamate by the brain tissue was approximately equivalent on a molecular basis with the K^+ uptake, and the investigators believe that glutamate plays a more intimate role in the transport itself than does the glucose, which probably acts simply as an energy substrate [14]. It is suggested that the cationic transfer is a response to the separation of hydrogen ions spatially, resulting

---

[14] However, in terms of K^{42} *exchange* rates, KOREY (1952) found no effect of glutamate or aspartate, or of pH, in rabbit cerebral cortical slices.

from operation of a hydrogen-carrier system (possibly the glutamate-iminoglutarate system) which depends on oxidative phosphorylation energy and in turn on glucose oxidation.

Resting State of Cations.—Rothenberg's (1950) studies with K^{42} and Na^{24} would at face value imply that, although the total Na^+ in the squid giant axon is readily exchangeable, a large part of the K is not in a free ionic form. This result is so contrary to the bulk of evidence on the matter, however, that it has not met general acceptance. Keynes and Lewis (1951) found at least 97% of the K in small bundles of *Carcinus* (crab) nerve fibres to be exchangeable. They observed however that the intracellular K was evidently in more than one compartment, as with K^{42} exit from the fibres into a balanced medium a progressively diminishing rate constant was seen; oddly enough, the loss into a K^+-free medium did not show this characteristic. The Q_{10} for this process was about 2.0. Hodgkin and Keynes (1953) similarly found that in *Sepia* (cuttlefish) giant fibres at least 90% of the axonal K was freely exchangeable; moreover, the mobility and diffusion coefficients of K^+ in the axoplasm were comparable to those in m/2 KCl.

This turnover can be very rapid; according to Krebs *et al.* (1951), during the steady maintenance state in guinea-pig brain slices, K^{42} turnover amounted to 3—4% exchange per minute. Fast as this rate is, it was still only about half that shown by pieces of ox retina. Keynes (1951 b) presents a very thorough analysis of the Na^+ and K^+ fluxes through the *Sepia* axonal membrane. In the medium used, the resting potential was somewhat lower than that calculated from the K^+ distribution, and accordingly K^+ leaked out along its electrochemical gradient, the outward/inward flux ratio being 3—4. The Na^+ influx was about double the efflux, but these rates were far lower than for K^+.

Relation to Neural Activity.—Keynes (1949) noted, as exemplified in Fig. 5, that the leakage of K^{42} from both *Carcinus* and *Sepia* axons was augmented considerably by conductile activity, as was the entry of Na^{24}, the total increment for each ion species being on the order of $5 . 10^{-12}$ mols/cm.^{2} of cell surface (about 30,000 ions/micron^2) per impulse. At the same colloquium, Hodgkin (1949) pointed out that these ion movements during conduction in the *Sepia* axon are of the right order to account for the observed voltage changes, in view of the measured electrical properties of the membrane. The resting potential is also accountable in terms of the K^+ gradient, on the basis of the classical assumption of a much higher permeability to K^+, at rest, than to Na^+ or intracellular anions.

But with the passage of the impulse, the picture is more complicated than in the older scheme. Hodgkin showed that if the NaCl in the medium was partly replaced by non-electrolytes or with choline chloride, or if extra NaCl was added (short of osmotic cellular damage), the effects on the amplitude and rate of rise of the action potential implied that the impulse involves a large transient increase in permeability to Na^+, as was indicated also by KEYNES's observations with Na²⁴. GRUNDFEST and NACHMANSOHN (1950) showed that squid giant axons also took in Na²⁴ at a markedly increased rate during repetitive firing, somewhat more so in the presence of cholinesterase inhibitors. KEYNES and LEWIS (1950) estimated by activation analysis that, following twenty minutes of continuous activity at ten impulses per second, the Na⁺ content of *Sepia* axons was tripled, while the K⁺ content fell to about ¾ of the original resting level. Combined with the tracer flux data, this would indicate that under these conditions, no

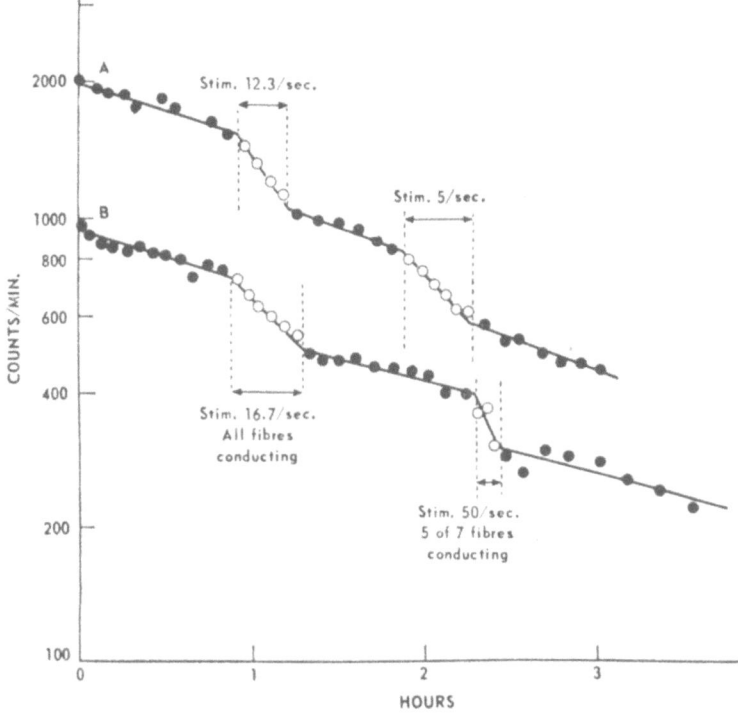

Fig. 5. Leakage of K⁴² from *Carcinus* nerve in normal Ringer's, as affected by stimulation.
A, whole nerve; B, a bundle of seven 30μ fibres. T = 16°C.
(Courtesy of R. D. KEYNES and Journal of Physiology)

large part of the immediate K⁺ loss had been recovered, whereas about half of the Na⁺ leaking in had been expelled. This aspect was amplified by KEYNES (1951 a): the loss of K⁴² from *Carcinus* fibre bundles conducting 50 impulses per second was about ten times the resting loss; but the influx measured with K⁴² under similar circumstances was only 1.7 times the resting level, confirming the notion that the net loss of K⁺ is a very large fraction of the efflux. It amounts to somewhat more than a thousandth of one percent of the fibre's K⁺ per impulse, which is more than enough to account for recharging of the depolarized membrane. The loss of K⁺ during activity and its recovery at rest were followed by SHANES (1951) in the leg nerves of another type of crab, *Libinia*, by analysis of very small volumes of sea water in contact with the nerve surface. KEYNES (1951 b) estimated that stimulation at 100/sec. increased the Na⁺ fluxes

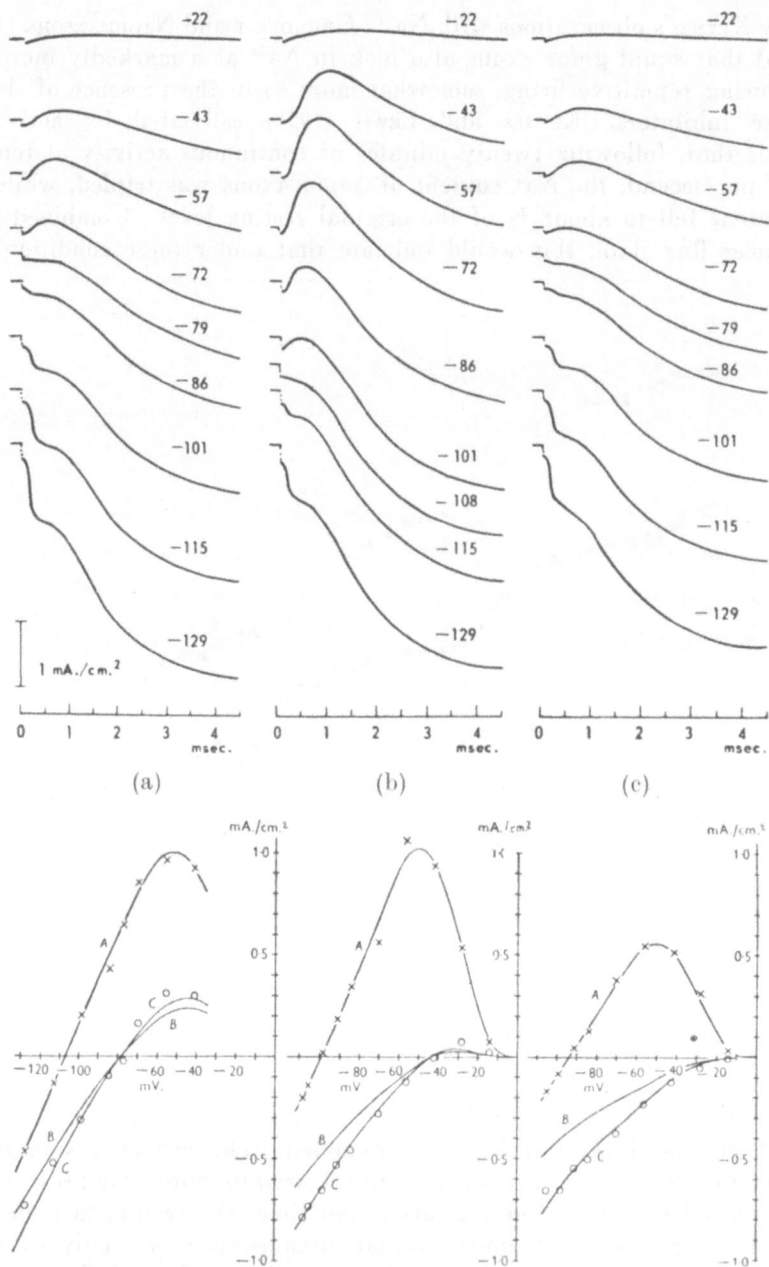

Fig. 6. Sodium current during axonal depolarization.   Upper records: curves of ionic current density during "voltage clamps". *a*, axon in 30% sodium sea water; *b*, axon in sea water; *c*, after replacing 30% sodium sea water. Numbers marking curves are fixed displacements of membrane potential, in millivolts. T = 6.3° C. Lower graphs: relation of peak Na+ current at fixed depolarization to external [Na+], in three experiments. Peak Na+ current observed during voltage clamp is plotted against maintained displacement of membrane potential. Crosses: axon in sea water; Circles: axon in low-Na+ sea water. Curve A drawn by eye through crosses, and used as basis to calculate Curve B for low-Na+ sea water, on basis of theory considered in text. Curve C represents Curve B with all ordinates multiplied by a constant factor, *f*.

(a) 6° C. Low-Na+ solution is 30% sodium sea water. *f* = 1.20. (b) 8.5° C. Low-Na+ solution is 10% sodium sea water. *f* = 1.333. (c) 8.5° C. Low-Na+ solution is choline sea water. *f* = 1.60.

(Courtesy of A. L. Hodgkin and Journal of Physiology)

about 20 times, tripled the K+ influx, and raised the K+ efflux by a factor of about 9.

The temporal characteristics of these flux changes accompanying membrane polarization changes were analysed in some detail by HODGKIN and HUXLEY (1952 a). Working with a section of squid giant axon in which the polarization was maintained by an electronic "voltage clamp" feedback system, they were able to record the actual membrane current with a fixed depolarization, as a function of the |Na+| of the medium (replacing any withdrawn Na+ by choline+). Fig. 6 illustrates the nature of the current responses recorded. The critical depolarization potential for abolishing the initial inward surge of current was found to vary with the [Na+] as would the potential of a Na-electrode (in Fig. 6, the theoretical shift of the x-intercept in Curves B and C corresponds to the observed shift indicated by the circles). This initial current, then, apparently results from a sudden rise of the Na+ mobility in the membrane, its magnitude and direction depending simply on the Na+ gradient and the set potential gradient (showing that it apparently involves no immediate active process). The delayed outward current, however, was not dependent on the external |Na+| and is probably carried mainly by K+. But when the membrane was repolarized to varying degrees by means of the voltage clamp (following a brief depolarization) the behavior of this potassium current as a function of [K+] showed that the potassium potential changes with [K+] in the expected manner qualitatively, but by only about half the amount expected for a K-electrode (HODGKIN and HUXLEY 1952 b). In summary, the evidence suggests that depolarization involves a rather abrupt rise in Na+ conductance which then declines exponentially, while the K+ conductance rises much more gradually and to a lesser degree, but remains elevated more or less indefinitely. HODGKIN and HUXLEY (1952 c) looked further into the nature of the drop in Na+ conductance during a maintained slight depolarization, and found that it involves a marked refractoriness, so that the conductance fails to show the expected transient increase upon further depolarization. In a similar arrangement, HODGKIN and HUXLEY (1953) followed with K^{42} the efflux from sections of *Sepia* axons as a function of the intensity of subthreshold cathodal currents, and found not only a direct proportionality, but an actual *equivalence* of the K+ loss and the quantity of electricity moved (1 mol per 96,500 coulombs), so that the K+ efflux must be the only significant component of the membrane current in such a partly depolarized axon.

We thus see that it seems possible, on the basis of present knowledge, to account for the cation behavior in these tissues, both at rest and during the conductile event, without invoking any driving forces other than the existing electrochemical gradients. The need for *active* transport in this picture (in the sense of moving ions such that the fluxes do not correspond to these gradients) comes in the *recovery* phase, after activity. At this time either Na+ must be expelled against the gradient, K+ accumulated against the gradient, or both, in order to reëstablish the original ionic distributions. In the *Sepia* axon, it appears (HODGKIN and KEYNES 1954) that the energy requirements for this could be met (but not many times over) at the observed metabolic rate. Ussing's coupled efflux-influx system seems inapplicable here, since the efflux of Na$^+$ persists in Na$^+$-free media. Furthermore, the efflux is reversibly inhibited by DNP, cyanide, or azide, with little effect on the influx. As in other tissues, the Na$^+$ efflux depends on the external [K$^+$], being sharply reduced in K$^+$-free media and enhanced somewhat by [K$^+$] above the normal level. Moreover, the K$^+$-influx inhibitor-sensitivity resembles that of the Na$^+$ efflux, while the K$^+$ efflux is not responsive to these agents. These facts recall the type of linkage of Na$^+$ and Ka$^+$ transports suggested for the red cell by Harris and Maizels (1952) and Harris (1954). On this basis, a large part of the normal influx must be by the secretory channel, and the match of the K$^+$ flux ratio with the electrochemical gradient in the unpoisoned *Sepia* axon (indicating a passive behavior) is merely fortuitous! This appears to be the present favored interpretation among the Cambridge group.

Cation Transport in Microörganisms

K$^+$ Movements and Carbohydrate Metabolism in Yeast.—In several cell types studied, and perhaps in all cells in which glycolysis is prominent, this activity involves a pattern of K$^+$ movements such as is well brought out in Pulver and Verzár's (1940 a, b) observations in baker's yeast. About $^1/_3$ of the K content of the yeast was readily washed out during preliminary exposure to a dilute saline medium; upon incubation with glucose at 20° C., K$^+$ was at first rapidly taken into the cells, and then began to reappear in the medium with the onset of CO$_2$ production. In the initial period, approximately one atom of K was taken up for each 4 molecules of glucose. No concurrent exchange of Na$^+$ could be demonstrated, and the timing of acid production was such that it was concluded that reciprocal K$^+$-H$^+$ movements were not significant in the process. As Table V illustrates, iodoacetate exaggerated the loss of K$^+$ and prevented both its uptake and the utilization of glucose; fluoride was effective only at much higher concentrations, while cyanide or phlorrhizin were not at all effective. That fermentative activity accelerated the exchange of K^{42} in both directions through the yeast cell surface was observed by Hevesy and Nielsen (1941).

Conway (1942) interpreted these findings as no evidence of a true specific transport system, but as simply another example of passive K$^+$ accumulation in response to the cell membrane's permeability limitations (as proposed for muscle cells by Boyle and Conway 1941), in this instance centering on the fixation of diffusible orthophosphates as indiffusible phosphate esters within the cell. This view was reasserted in the demonstration (Conway and O'Malley 1943) that NH$_4^+$ could substitute for K$^+$ in accompanying the apparent phosphate movements. Later however (Conway and O'Malley 1944) the K$^+$ movements were attributed to an exchange for H$^+$ from some metabolic product, probably pyruvate; since the cell mash was, if anything, left more alkaline by fermentation, it was

Table V. *Potassium Movements Accompanying Glucose Assimilation by Yeast.*
(From PULVER and VERZÁR, 1940 a.)

| Time (min.) | K or Glucose in Supernatant Fluid (mg./5 ml.) | | | | | |
| | No Inhibitor | | Iodoacetate, M/2000 | | Fluoride, M/300 | |
	K	Glucose	K	Glucose	K	Glucose
0	0.672	33.0	0.686	31.5	0.677	33.1
10	0.196	3.75	0.852	26.2	0.625	29.3
20	0.288	1.5	0.953	18.1	.0.562	20.0
40	0.352	—	1.087	10.8	0.495	11.3
60	0.571	—	1.317	8.7	0.855	1.7
120	0.828	—	1.773	6.2	1.505	—
Start of CO_2 production	6′		8′		8′	
Amt. of CO_2 produced during period 10′–15′	16.2 cc.		1.1 cc.		1.35 cc.	

Yeast in 20% suspension; glucose added at 0 min.

suggested that only a limited outer region of the cell was involved in the H^+-K^+ exchange [15].

ROTHSTEIN and ENNS (1946) established and quantitated such an exchange. When Na^+ was the only cation available in the medium, succinic acid left the slowly fermenting cells; supplying K^+ permitted its exchange for the H^+ from such acids, and an increased rate of fermentation. When the substrate was consumed, some K^+ leaked out again, but an amount determined by the deposition of polysaccharides remained inside to be released only with dissimilation of the carbohydrate. O_2 lack approximately halved both the K^+ retention and the assimilation; azide or DNP inhibited both processes, but still permitted rapid glucose catabolism. CONWAY and O'MALLEY (1946) figured that about 80% of the K^+ movement is in direct exchange for H^+, the rest involving movement of Cl^- along with the K^+. They noted that if certain complications regarding fixation of K^+ within the cell are postulated, the whole system can be interpreted on the passive BOYLE-CONWAY basis.

CONWAY and MOORE (1950) observed however that the addition of azide after the initial K^+-H^+ exchange produced a rapid reversal (loss of K^+ and uptake of H^+), although none of the lost succinate was recovered. In K^{42} tracer experiments, the K^+ lost in the presence of azide showed the same specific activity as the total cell K, so that no special intracellular partition-

---

[15] CONWAY and DOWNEY (1950) present more direct evidence that about 10% of the yeast cell volume, near the surface, is in a zone which is differentiated permeability-wise from the cell interior.

ing was evident. In comparing the effects of an extensive array of sub-strates, Ørskov (1950) found a definite parallelism between the K+ uptake and the O_2 consumption.

(Nickerson and Zerahn, 1949, showed that yeast could accumulate Co^{60} to levels much higher than in the medium, but the poor exchangeability and the ex-tractibility properties of the cell-Co suggested that it was combined rather rigidly to the cell surface, so that probably no real *transfer* was indicated here.)

Metabolic Acquisition of K+ in Bacteria.—*Escherichia coli*, like yeast, incorporate K+ when glucose is first added to the medium and begin to lose it again while the glucose consumption is still going on, as shown by Leibowitz and Kupermintz (1942). In their experiments, the K+-uptake and fermentation rates were affected in a parallel manner by the temperature, pH, substrate concentration, and oxygen supply. The initial sugar uptake far exceeded the fermentation, and the K+ retention was associated with this excess which was in the form of a non-reducing, hydrolysable poly-saccharide. Ørskov (1948) reported that doubling the tonicity of the medium induced a considerable uptake of K+ in these bacilli; and that this effect was blocked by azide, iodoacetate, or cyanide, but not by fluoride.

Cowie, Roberts, and Roberts (1949) supplied the basic study on Na+ and K+ turnover rates in *E. coli*, showing a very rapid equilibration of isotopic tracers for both cations, even in resting cells. The Na+ distribution appeared to be dictated by diffusion only and was unresponsive to the metabolic state of the cell, whereas activity and growth involved an apparent binding of a considerable amount of K. Roberts, Roberts, and Cowie (1949) further noted that the uptake of K+ following the addition of glucose was an abrupt one, the level stabilizing after the first few minutes; this level was increased with the levels of glucose and K+ in the medium up to a limit of about 2 g. of K per liter of cells. The pH and the supply of O_2 and phosphate did not appreciably affect the K+ uptake, but aeration did accelerate the later loss of K+. Both phases of the ex-change were depressed by DNP, azide, or iodoacetate. Glucose-1-phosphate (G-1-P) was much more active than glucose in effecting the K+ uptake: and fructose, galactose, ribose, glycerol and glutamate were all effective to some degree. Rb+ and K+ competed in this system as elsewhere. A similar process was described also in the flagellate *Chlamydomonas*, in the bacterium *Rhodopseudomonas*, and in the ovarian eggs of the salamander.

A series of reports from Oxford (Eddy and Hinshelwood 1950, 1951: Carroll *et al.* 1950; Eddy, Carroll *et al.* 1951) describe a similar K+ ac-cumulation associated with metabolic activity in *Bact. lactis aerogenes*. Rb+ competes in this system with 30–40% of the effectiveness of K+, but again neither Na+ nor Li+ nor Cs+ can take part. But tracer studies showed that the retention in this instance does not involve a rigid binding of K+: moreover, the intracellular K+ readily exchanges with H+ at low pH. Rapid uptake of K+ or Rb+ was initiated by presentation of glucose. pyruvate, succinate, or other substrates, regardless of whether a source of N was provided so as to permit actual growth (Fig. 7). Associated with this temporary K+ retention is a parallel phosphate accumulation: some type of coupling of the two phenomena is clearly suggested, but the experiments do not provide specific information on this point.

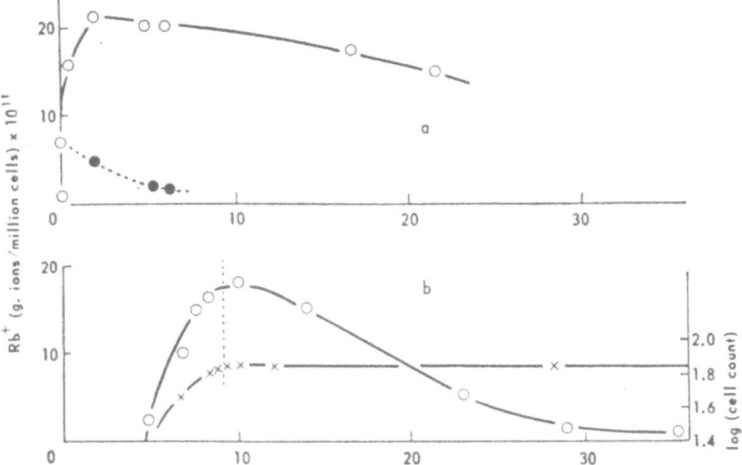

Fig. 7. Uptake of Rb+ during glucose oxidation in *Bact. lactis aerogenes*. *a*. Without growth (no N-source): abscissa is hours from time of suspension. Open circles: glucose present; solid dots: no glucose. *b*. With growth; abscissa is hours from time of inoculation. Crosses: growth curve; circles: Rb+ uptake.

(Courtesy of Sir CYRIL HINSHELWOOD and the Royal Society)

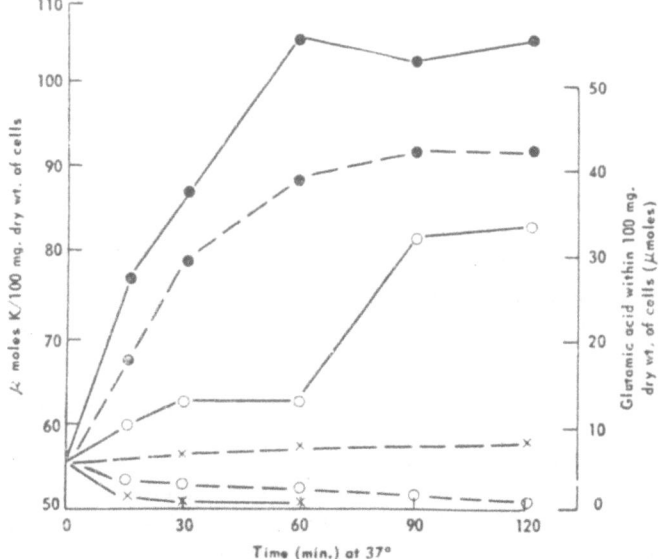

Fig. 8. Accumulation of K and glutamic acid by washed *Staph. aureus*. Cells incubated (5 mg. dry weight per ml.) for 1 hr. in minimal-K+ solution at pH 5.0, with or without 2% glucose and 5 micromolar glutamic acid. Solid lines — K+; broken lines — glutamic acid. Open circles — with glucose only; crosses — with glutamic acid only; solid dots — with both.

(Courtesy of Biochemical Journal)

DAVIES *et al.* (1953), studying primarily the uptake of amino acids by microörganisms (which will be taken up in a later section) observed that this process often involves an accompanying shift in cations. Fig. 8 shows

for example the entrance of K+ accompanying the glucose-stimulated uptake of glutamic acid by *Staphylococcus aureus*. Note that glucose in
itself also initiates a lesser degree of K+ uptake. These phenomena were
observed also in *Streptococcus faecalis* and *Saccharomyces fragilis*; the
incremental K+ entry never exceeded the amino acid uptake, mol for mol.
The yeast also took in Na+ as well as K+, and *Staphylococcus* would do
so only in the absence of K+. The streptococcal uptake of lysine also
involves K+ gain, whereas the same process in the staphylococcus does not,

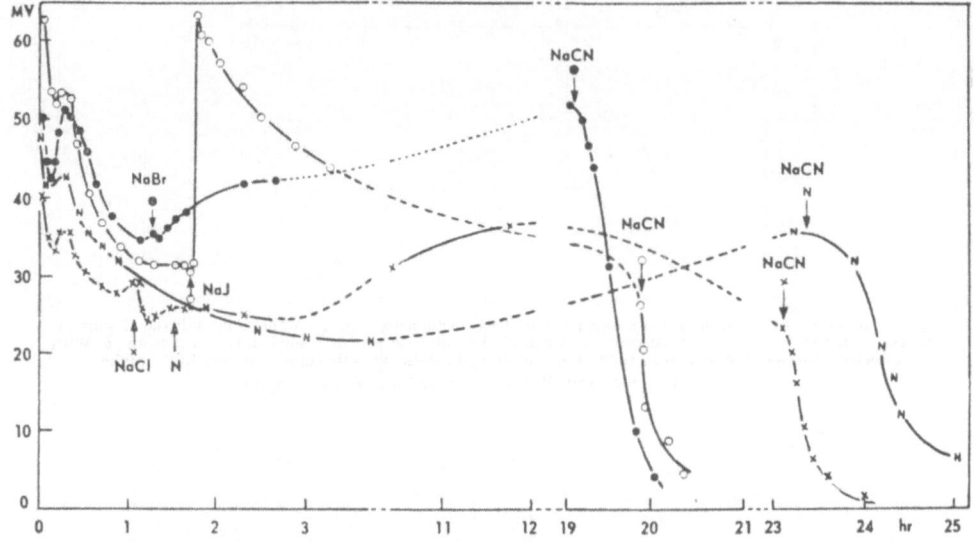

Fig. 9. Influence of various anions on the potential difference across frog skin. 0.4 Ringer's on both sides
of the skin at the beginning of the experiments (everted leg-skin bags). At arrows, salts as indicated were
added to outside solutions, raising Na+ level from 48 mM to 68 mM. Record **N-N-N** is control in which
no further salt was added. All experiments were terminated by addition to outer solution of small amount
of NaCN.

(Courtesy of E. G. Huf and Archiv für die gesamte Physiologie)

and in the yeast it actually entails a *loss* of Na+ and K+. Thus no clear
pattern associating the uptake of the two classes of substances can be
brought out.

Cation Transport through Amphibian Skin

Properties of Salt-Intake System.—Frogs which have been depleted of
salt by washing with distilled water will absorb NaCl through the skin
even from most dilute solutions, but when exposed to KCl, NH₄Cl or CaCl₂
will for the most part merely exchange HCO_3^- for Cl- (Krogh 1937 a).
NaBr is also taken up rather effectively, but NaI almost not at all. Krogh
therefore tentatively concluded that the Cl- is the actively transported
ion here, but allowed for the possibility also of an active Na+ transport,
which has emerged in later studies as the primary process.

Huf and his associates have made extensive use of the skin peeled from
frog legs, in the form of bags holding the external solution (everted bag)
or the internal solution (bag right-side-out). The bag contents are analysed

after 12–24 hours' immersion in the experimental medium. In the earlier group of studies in Germany (HUF 1935 a, b; 1936 a, b, c), some very old observations were verified regarding the movement of NaCl and water from outer to inner compartments, beginning with equal tonicities of Ringer's solution on each side. Cyanide or bromoacetate inhibited this process and markedly depressed the skin p. d. (Fig. 9); lactate or pyruvate overcame the bromoacetate-poisoning, but glucose was not very effective here. All of this paralleled the responses in the oxygen consumption rate of the skin, so that it was suggested that lactate-pyruvate oxydation energy was involved in the salt transfer process. KATZIN (1940) showed that also in terms of Na²⁴ exchange the "permeability" inward for Na⁺ might be as much as four times the outward permeability. More recently HUF's work has been resumed in Virginia; it was shown that the net rate of NaCl uptake (extrusion from the everted bag) was raised by increasing the external concentration, but higher *gradients* were maintained with the more diluted Ringer's solution (HUF, PARRISH, and WEATHERFORD 1951). Tem-

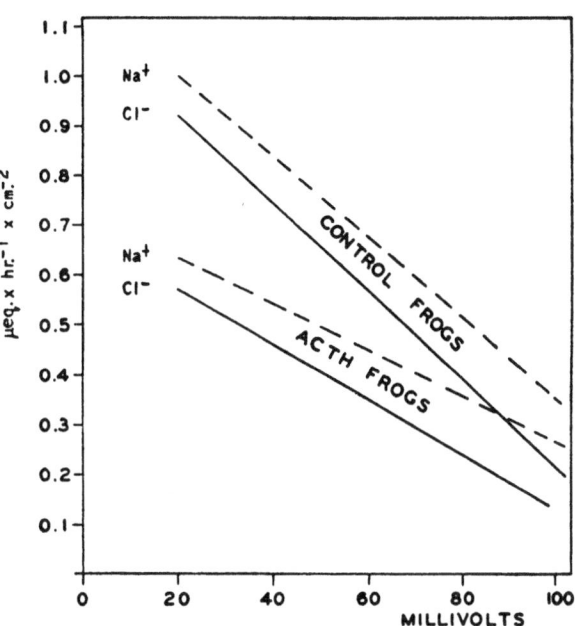

Fig. 10. Relation of Na⁺ and Cl⁻ uptake to frog skin potential. Isolated pieces of *Rana pipiens* skin. The "ACTH frogs" had been pretreated by implantation with a purified ACTH preparation. The four linear functions were obtained as best rectilinear fits of fairly scattered data in the four independent experiments.
(Courtesy of E. G. HUF and Journal of General Physiology)

perature had little effect on the concentration ratio established. A small amount of K⁺ was essential for the operation of the pump (HUF and WILLS 1951); Rb⁺ could replace K⁺ in this respect, Cs⁺ being a less effective substitute, and NH₄⁺ being worse than nothing at all. The Mg⁺⁺ and Ca⁺⁺ levels appeared to be immaterial.

Cl⁻ movement evidently does not quite keep pace with the Na⁺ intake (see Fig. 10), the external levels of K⁺ and H⁺ rising somewhat by exchange for the entering Na⁺. According to HUF, WILLS, and COOLEY (1951) and HUF and PARRISH (1951), replacement of part of the Cl⁻ by Br⁻, NO₃⁻, HPO₄⁻⁻, SO₄⁻⁻, glycerophosphate, or ATP did not markedly alter the Na⁺ uptake at a given [Na⁺], but I⁻ had a distinct depressing effect. Accompanying this depression was an increase in K⁺ exchange in the opposite direction and increase in the electrical polarization of the skin (the interior being positive to the exterior as illustrated in Fig. 9). The situation is thus apparently one of active uptake of Na⁺ specifically, the anions following by reason of electrical forces, I⁻ having a notably lesser mobility in the barrier than the other anions.   In comparison of individual specimens (Huf and Wills 1953) a definite positive correlation was noted

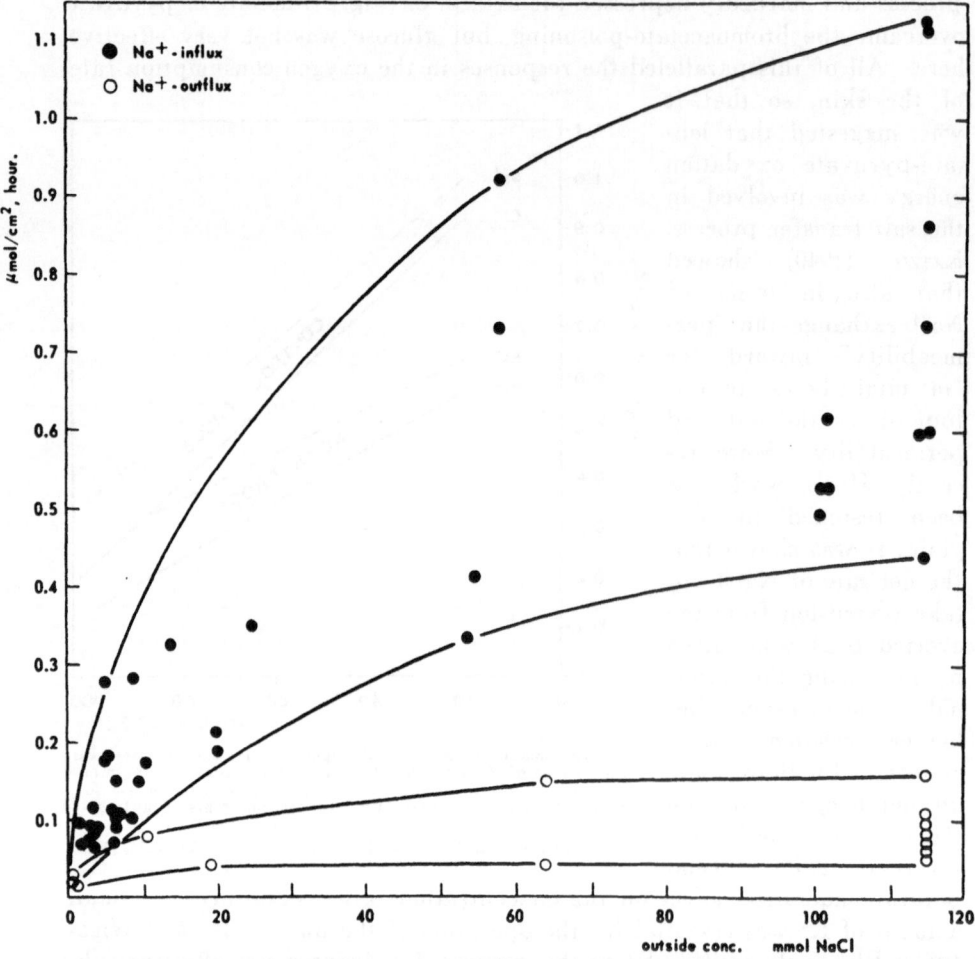

Fig. 11. Na⁺ influx and outflux through isolated frog skin as functions of the outside NaCl concentration. Inside solution is isotonic and at pH 8.1- 8.3. Outside pH near neutrality. Flux estimated from Na³⁴ transfer over 1–2 hour periods.
(Courtesy of H. H. Ussing and Acta Physiologica Scandinavica)

between the skin p. d. and the K⁺ output; but the p. d. fell with increasing NaCl uptake, as shown in Fig. 10. This is exactly what would be expected on the basis of an active Na⁺ transport with the K⁺ and anions moving passively, if the individual skin specimens varied in their electrical conductance.

Levinsky and Sawyer (1953), using right-side-out skin bags, noted inhibition of NaCl accumulation by DNP at 10⁻⁵ M, suggesting the importance of oxidative phosphorylation energy in the transport; the degree of inhibition was linear with the log [DNP]. This agent reduced the loss of H^+ from the interior, and reversed the normal K^+ exit; however, the extra K^+ found inside the DNP-poisoned bags resulted from a loss from the skin tissue itself. A variety of other inhibitors produced a similar effect.

UsSING and his associates at Copenhagen have provided the most thorough analysis of the Na^+-transport system in the frog skin; this group

Table VI. *Influence of Various Factors on Na^+ Influx through Isolated Frog Skin, as Traced with Na^{24}.*

(Adapted from data in UsSING, 1949.)

Conditions and Agents Applied	Na^+ influx (mμM/cm.^{2}-hr.)
Outside pH 6.5; inside pH 6.72	18, 10
7.77	60, 61
Outside pH 6.5; inside pH 8.1	318, 267
7.2	161, 70
Inside pH 8.0; outside pH 6.6	105
5.25	97
3.3	12
No special agent	92
Adrenalin on inside, 1 : 10^6	185
Adrenalin removed	141
No special agent	121, 118[1]
Cyanide on inside, about 1.6 mM	72, 18, 28[1]

[1] values in last experiment are approximated from a graph.

Multiple entries in table are for successive hours; all testing periods were of one hour duration.

Outside salt concentration was 115 mM in all; inside concentration varied over range, 3–6 mM.

generally uses a section of *Rana temporaria* abdominal skin separating two salt solutions in a special chamber which permits certain electrical measures and controls (to be mentioned later) while the fluxes are followed with radioisotope tracers [16]. UsSING (1948, 1949) found that the Na^+ influx

[16] UsSING (1948) brings out a valuable admonition in regard to quantitative estimation of active transport from tracer flux data. In a system operating by exchange diffusion, or where there are opposed electrical and chemical gradients, these methods may seem to imply an active transport where none exists, or may exaggerate its contribution to the whole picture; on the other hand, the same methods are likely to underevaluate the activity of a transport system if there is more than one barrier offering a high resistance to diffusion between the compartments analysed, since the internal back-losses will then not be accounted for in the end-to-end fluxes.

always exceeded the efflux even if the outside level were very low; as Fig. 11 shows, both fluxes rose as the external [Na+] was raised, but not linearly, the uptake showing "saturation kinetics." Cyanide-poisoning nearly abolished the influx (and reversed the normal p. d. across the skin), while the outflux continued unabated. The net transport was rather insensitive to the external pH, but *internal* alkalinity distinctly enhanced the NaCl accumulation, accelerating influx without altering efflux. Similarly, adrenalin was ineffective on the outside, but at 10^{-6} on the inner surface approximately doubled the influx while producing an enormous transient increase in the leakage outward with a concomitant drop in the p. d. (see also Jørgensen 1947, concerning this). Cyanide prevented also this response to epinephrine. Table VI illustrates several of these points.

In summary, these observations point clearly to the influx process as the active phase, and to the inner side of the skin as the site of the activity. Histologically, the only likely layer in the skin for the secretory activity is the innermost layer, the stratum germinativum, and by reason of the facts just reviewed, Ussing feels that the active Na+ transport occurs as an extrusion through the proximal cell membranes of this layer into the interstitial fluid. He suggests that the control of this process by the pH on the inner side involves the cell's adjustment to maintain its own internal pH by exchange of Na+ for H+, the latter being the most freely penetrating cation available when the Na+ is extruded from the cell. The fact that increased P_{CO_2} rapidly depresses Na+ influx, while with phosphate buffering the pH effects show some latency, reinforces this notion that the *intra*cellular pH may be the key factor.

The Unique Role of Na+.—The possibility of involvement of Cl− in the direct "pump" action is perhaps not definitely precluded, but concentration and potential measurements do not point to any deviation from thermodynamic equilibrium in the distribution of this ion. Moreover, flux measurements with Cl^{36} (Koefoed-Johnsen, Levi, and Ussing 1952) show a flux ratio equal to the electrochemical activity ratio, with isolated pieces of the skin. By use of K^{42} tracer, Levi and Ussing (1949) showed that with higher skin potentials, the flux ratios for K+ (when [K+] was 3.5 mM on each side) were also in reasonable accord with the notion that K+ moved through only passively. However, with less polarized skin, the influx exceeded the efflux (opposing the potential gradient), so that some degree of activity must be involved in the influx. Ussing and Zerahn (1951) present a beautifully conceived simple demonstration that the Na+ transport is the only significant active factor in the system and is, in fact, the source of the skin p. d. This involved passing through the skin an electric current from an external battery, such that the potential drop across the skin was nil; with identical solutions on both sides, no net transfer of passive ions could then take place, while actively transported ions would show a net movement. In this arrangement, the Na+ influx (determined

Table VII. *Na+ Transport in Short-circuited Frog Skin.*
(Modified from Ussing and Zerahn, 1951.)

Treatment	Influx Expts. mCoulombs/cm.²-hr.		Outflux Expts. mCoulombs/cm.²-hr.	
	Na-flux	Current	Na-flux	Current
Control	57	49	13.6	112
Adrenaline on inside, at about 1 : 100,000	87	76	41.0	126
	111	88		
Control	126	118	1.6	77
Whale neurohypophysealextract,				
on inside	246	232	3.1	129
(1 mg. dry gland/40 ml.)			5.4	164
Under 5% CO_2 : 95% O_2	4.5	0	5.5	0
	3.5	0	6.1	0
Under atmospheric air	165	150	8.3	161
	173	163	15.3	158

Figures in individual experiments are for successive one-hour runs.

with Na^{24}) proved always to exceed slightly the total current, as shown by the examples in Table VII. The Na+ efflux concurrently determined with Na^{22} just about accounted for the difference between the two figures, so that the entire short-circuit current was attributable to active Na+ transport. This experimental system further permitted the calculation of the electromotive force and internal resistance of the "sodium battery." Linderholm (1952), using similar methods in Teorell's laboratory, showed that the combined conductances for Na+ and Cl− account for the electrically estimated total conductance for the frog skin. Linderholm's (1953) demonstration that the p. d. does not reflect the Na-pump potential or the Na conductance, but varies inversely with the Cl conductance, underscores the roles of the two ions in the picture.

Ussing (1954) notes that Li+ competes with Na+ with approximately equal propensity for the carrier system in mixtures up to 80 Li+ : 20 Na+; however, 100% Li+-Ringer's solution is toxic. This summary report of Ussing's presents a thoughtful and comprehensive review of the work of his group.

Action of Hormonal and Other Agents.—The actions of various inhibitors have been considered in developing the arguments above. Ussing and Zerahn (1951) looked further into the effects of various agents acting on the short-circuited skin system (Table VII). Cu++ elevated the p. d. without altering the Na+ flux or current, presumably by slightly depressing the passive Cl− penetration; adrenalin and neurohypophyseal factors (from the whale) enhanced the flux and current considerably, and in parallel [17]; high CO_2 abolished the influx and current. In the examples in the table, no involvement of any ion other than Na^+ in the active transfer need be suggested, except for the case of adrenalin stimulation. Here it has been shown that another active factor enters the picture. With this agent, the short-circuited skin current may nearly double and this increment appears to be essentially an outward Cl^- transport, as Koefoed-Johnsen, Ussing, and Zerahn (1952) showed with simultaneous determinations of influx and outflux with Cl^{36} and Cl^{35}. But this probably is by way of a mucus gland secretion.

Jørgensen, Levi, and Ussing (1947) examined the action of neurohypophyseal principles (from the ox) on intact axolotls in a circulating system of very dilute, labelled NaCl. Injection of the pressor principle induced a rather protracted net uptake of salt while the oxytocic principle caused a rapid and prolonged loss. Koefoed-Johnsen and Ussing (1949) found in the same preparation a large progressive uptake of Cl^- following injection of a corticotropin preparation from the beef pituitary; the Na^{24} influx was doubled or tripled for several days by this treatment. On the other hand, as noted incidentally in Fig. 10, several days' pretreatment with ACTH considerably reduced the NaCl uptake of the isolated skin, at a given skin potential, in the frogs used by Huf and Wills (1953). The cortical hormones, however, did not produce this change.

Fuhrman (1952) investigated the effects of a large group of metabolic poisons on the Na^+-transport system, measuring also the Cl^- flux with Cl^{36}, and the O_2-consumption of the skin. All effective inhibitors of respiration depressed the Na^+ influx, but some of the transport inhibitors (such as DNP) enhanced the respiratory rate, and many of them increased the passive Cl^- fluxes. Kirschner (1953) reported that anticholinesterases, applied to the inside of the skin, nearly abolished the Na^+ transport after some delay; eserine acted reversibly, TEPP [18] irreversibly. At the same time, the D. C. resistance was markedly increased. On the outside surface, eserine was not very effective. Atropine on the outside reversibly and markedly elevated the Na^+ accumulation in *Rana esculenta*, but oddly enough this effect was not seen in *R. temporaria*.

Linderholm (1952) made excellent use of this frog skin Na^+-transport system to elucidate the basic action of some of the chief clinical diuretic agents, and confirmed thereby many of Ussing's observations. He argues that Conway's (1951) "redox pump," as the basis for this Na^+ transport, provides the most complete interpretation of the experimental facts. Linderholm also noted that, although perhaps only 5% of the energy of the skin's respiration is required for the Na^+ pump, the Na^+ *charge* transfer is 70–90% of the electron transfer involved on the basis of Conway's hypothesis; thus the "coulomb efficiency" appears to be very high indeed.

---

[17] The active pituitary principle appeared to follow the oxytocic factors rather than the antidiuretic and pressor factors (Fuhrman and Ussing 1951).

[18] Tetraethylpyrophosphate.

Renal Tubular Cation Transport

Na+ and K+ Transfer between Blood and Tubular Lumen.—RAMSAY (1951) found that the mosquito larva in a fresh-water environment could reduce its output of Na+ in the Malpighian tubules so that the "urine" was hypotonic to the haemolymph, but that it could not eliminate excess Na+ by this route when in a hypertonic environment. Examining the K+ level in the fluid in the Malpighian tubules of eight species of insects (in five different orders), RAMSAY (1953) noted that it always exceeded the level in the haemolymph, while the Na+ gradient was in the opposite direction. Considering the p. d. across the tubular wall, he took this as evidence of an active extrusion of K+ into the tubules.

In the mammalian kidney, although the gross fact of reabsorption of salts from the tubule lumen has been appreciated for many years, it was only recently that the question of a cation-transport system here has been given direct attention. BALDWIN *et al.* (1950) showed in dogs, in which small plasma upsets in [K+] and [Na+] were maintained by intravenous infusion, that the clearance behavior of both these cations indicated an active reabsorption of definite capacity. Although high levels of either cation in the plasma seemed to depress the tubular reabsorption of the other, the data did not imply an actual competition between Na+ and K+. Infusing LiCl in dogs greatly enhanced the K+ clearance (FOULKS *et al.* 1952), but since Li+ showed an entirely passive behavior in its own movements, the nature of its effect on the renal handling of K+ is not understood.

The observations of MUDGE *et al.* (1949) on Na+ excretion in dogs with severe urea diuresis showed that this ion is reabsorbed even against very considerable concentration gradients. Mercurial diuretics depressed the absolute rate of reabsorption, but did not alter the concentration ratios that could be maintained across the tubular wall.

PITTS *et al.* (1948) concluded from extensive observations in acidotic humans that the urinary acidification is largely accomplished by exchange of H+ for the Na+ reabsorbed in the tubule; the filtered buffer ions could not in themselves account for the quantity of acid eliminated.

In addition to reabsorbing filtered ions, the tubules may also actively secrete (excrete) cations into the lumen. MUDGE *et al.* (1948) observed secretory elimination of K+ in dogs undergoing urea diuresis, and BERLINER and KENNEDY (1948) established that in dogs given reasonable K+ loads, more than half of the elimination of K+ by the kidneys may be by way of active excretion through the tubules. Similar behavior was seen in human patients with disorders leading to elevated K+ excretion (BERLINER, KENNEDY, and HILTON 1950). MUDGE *et al.* (1950) noted that this system seemed to be called into play in dogs according to the state of the body's hydration, rather than in response to alterations specifically in the K+ level or its rate of filtration through the glomeruli. Thus reabsorption of K+ was prominent during water diuresis, while its tubular secretion was most evident in cases of extreme cellular dehydration. The secretion was depressed by the mercurial diuretics. According to BERLINER, KENNEDY, and ORLOFF (1951), mercurials also prevent the augmented K^+ excretion seen in dogs treated with 6063 (a potent inhibitor of carbonic anhydrase) so that this extra excretion is probably also attributable to tubular secretion. The fact that the drug prevents normal acidification of the urine in the face of a severe acidosis was taken as evidence that K^+ and H^+ may compete for a common secretory mechanism. Stamler (1951) showed, however, that the excretion of K^+ or Na^+ by dogs was not disturbed by saturating various other tubular transport systems, both excretory and reabsorptive.

The double treatment given to K^+ in the kidney tubule probably involves different loci. In a recent review of the subject, Mudge (1954) calls attention to the micropuncture work of Bott on the *Necturus* kidney, which shows directly that the filtered K^+ is actually nearly completely reabsorbed in the proximal tubule, only to be replaced by active tubular secretion in the distal tubule.

Cell-K Maintenance.—According to Conway, Fitzgerald, and Mac-Dougald (1946), the proximal tubular cells of the isolated frog kidney can accumulate very high concentrations of K^+ when an excess of this ion is supplied in the medium; but these cells are essentially impermeable to Na^+, thus falling into the Boyle-Conway pattern as originally developed for muscle cells. On the other hand, the *distal* tubule cells did not take up K^+ readily, but admitted Na^+ which was then apparently actively extruded.

Like other tissues when incubated in the usual balanced salt media, kidney cortex slices at first lose K^+ precipitously and then gradually recover the major part of the lost K^+ if metabolism is maintained. In such studies, Mudge (1951 a, b) generally exposed rabbit kidney slices first to isotonic NaCl for a few hours, so that about half of the original K^+ was leached out and replaced by Na^+ at the beginning of the experimental period. He found that under such circumstances the K^+ reaccumulation was blocked by anoxia, low temperatures, azide, cyanide, or arsenite, but was relatively insensitive to pH, osmotic pressure, and the common anions and divalent cations of biological media. Figs. 12 and 13 give representative results. The recovery of K^+ was roughly proportional to its level in the medium, up to about 80 mM. Addition of acetate (Fig. 12) added a fairly constant increment at all $[K^+]$; essentially all the Krebs-cycle ingredients similarly enhanced the K^+ uptake, but as in liver, glucose was ineffective. The action of all these substrates and inhibitors on the oxygen consumption paralleled their effects on the K^+-transport system. However, as shown in Fig. 13, Hg^{++}, Cu^{++}, and benemid interfered with accumulation at concentrations not depressing the respiratory rate; and DNP and some of its relatives depressed the K^+ uptake even while stimulating respiration. Involvement of oxidative phosphorylation energy in the process was implied by the correlation of phosphorylation-uncoupling effectiveness of the several related inhibitors and their inhibitory action on the potassium transport. The system was also depressed by fluoride, bromoacetate, iodoacetate (but not fluoroacetate), and somewhat by phlorrhizin; but there was no effect with inhibitors of carbonic anhydrase, cholinesterase, or alkaline phosphatase. Concurrent uptake of PSP, PAH [19], or diodrast did not alter the rate of K+ accumulation.

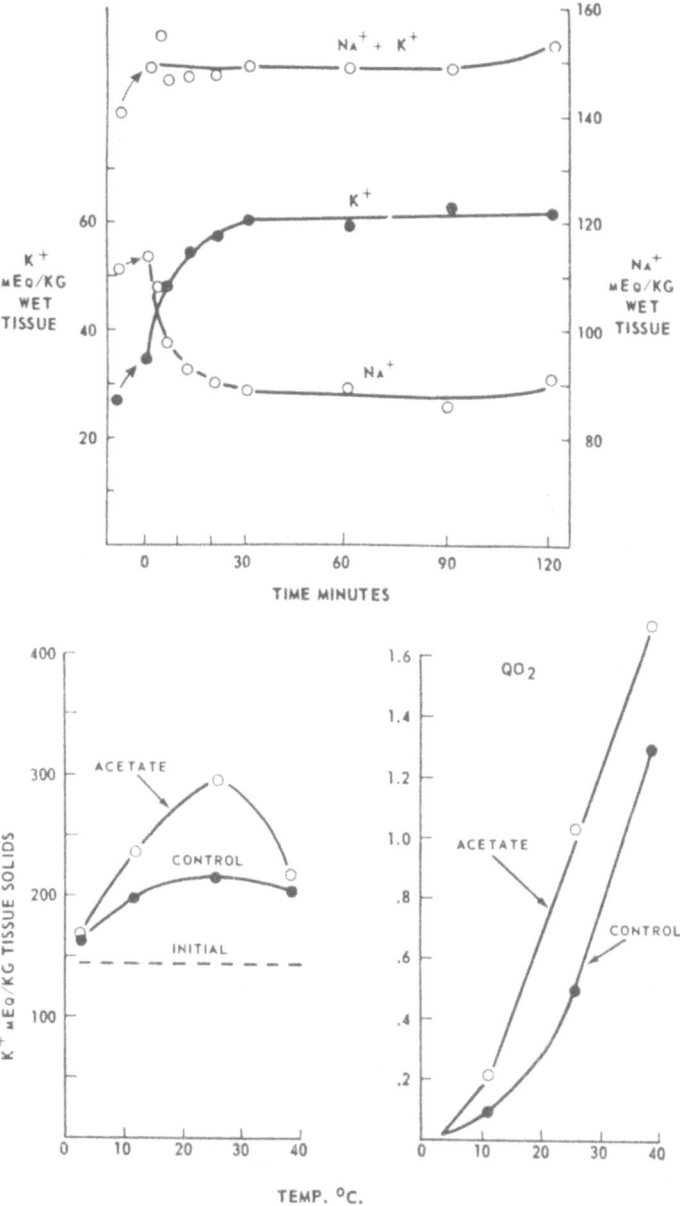

Fig. 12. Above: K accumulation with reciprocal changes in Na in rabbit kidney slices. Slices leached 2–3 hrs. at room T in 150 mM NaCl, then placed at 25°C. in acetate medium; KCl added, to 10 mM level, at zero time, after 10 min. of equilibration. Below: Effect of acetate and temperature on the process. K-levels plotted are 30-minute levels, in experiments like those of graph above.

(Courtesy of G. H. MUDGE and American Journal of Physiology)

[19] Phenolsulfonphthalein and p-aminohippurate, substances actively secreted by the kidney tubules and commonly used in kidney function tests.

This same system has also been tackled by the Sheffield group; Whittam and Davies (1953 a, c) noted in these slices that the Na+ content rose rapidly, replacing the lost K+, if the metabolism was suppressed by cold or by lack of oxygen (Table VIII). Under these circumstances, the α-ketoglutarate ratio (inside/outside) was only about $^1/_3$. With oxygenation

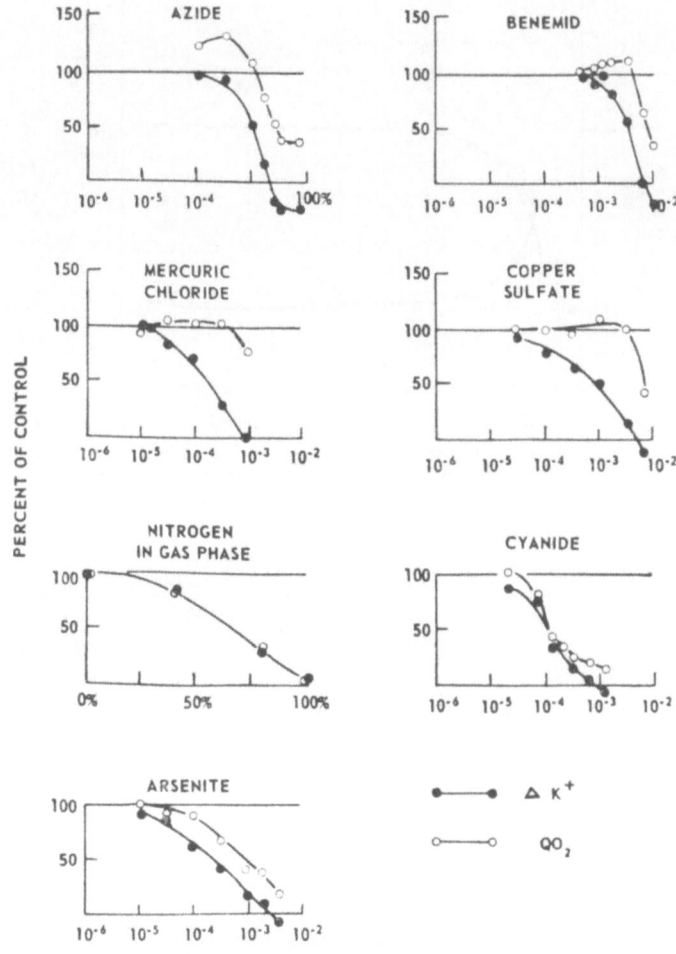

Fig. 13. Effect of metabolic inhibitors on K+ accumulation and respiration, as in Fig. 12. In all experiments, acetate was provided at 10 mM.
(Courtesy of G. H. Mudge and American Journal of Physiology)

at 37° C., as shown in the table, the ionic upsets were far less severe, and an intracellular ionic excess was maintained; this excess was just about equivalent to an additional fixation of α-ketoglutarate which was concentrated intracellularly by a factor of 2.5.

Aebi (1953) compared a variety of mammalian tissue slices in this type of experiment, and found the kidney by far the most active in accumulat-

Table VIII. *Metabolism and Maintenance of Water and Electrolyte Distribution in Guinea-Pig Kidney-Cortex Slices.*
(From WHITTAM and DAVIES, 1953 c.)

Conditions of Incubation	% H$_2$O		µM/g. of Na+		µM/g. of K+		µM/g. Total		Int/Ext. Conc. Ratio
	Before	After	Before	After	Before	After	Before	After	
Aerobic 0° C.	77.7	83.0	63.4	112	76.4	30.9	139.8	142.9	1.07
37° C.		79.7		90.2		75.3		165.5	1.29
Anaerobic 0° C.		83.5		109		32.2		141.2	1.05
37° C.		82.7		114		27.5		141.5	1.06

Incubation for 30–50 minutes in bicarbonate-saline with α-ketogluterate. Figures are averages from several experiments.

ing electrolytes during active metabolism. This phenomenon will be considered later in connection with active water transport and osmoregulation in mammalian tissues.

The Steady Cationic Exchange.—The study of K^{42} exchange (MUDGE 1953) gave further insight into the picture. Aerobically, the kinetics indicated at least two intracellular fractions of differing exchangeability, but even the slower compartment had a half-time of only about 25 minutes. However, with O$_2$ deprivation or DNP treatment, a significant fraction of the kidney-slice K became practically non-exchangeable. The magnitude of this non-exchangeable fraction was increased by Hg++, 6.10^{-4} M. Iodoacetate had no effect on the exchange, although as noted above it depressed the transport. WHITTAM and DAVIES (1954) do not agree that DNP induces the appearance of an unexchangeable fraction of tissue-K.

DAVIES and GALSTON (1951) found an extremely rapid K^{42} turnover (about 15% per minute) in guinea-pig kidney-cortex slices during the steady-state period following addition of α-ketoglutarate at 10^{-2} M (when no net accumulation or loss is occurring); this turnover is twice as fast even as that seen in the ox retina. According to WHITTAM and DAVIES (1953 b), the K^{42} exchange aerobically has a higher temperature coefficient than does the anaerobic transfer.

The somewhat faster Na^{24} turnover also shows two or more compartments for the cellular Na (WHITTAM and DAVIES 1954). Although MUDGE (1953) noted that the Na^{24} transfer did not show the signs of metabolic dependence exhibited in the K^{42} experiments, WHITTAM and DAVIES (1953 b) calculated that only a fraction of the Na+ distribution could be by simple passive means.

Na+ and K+ Transfer in Other Cell Types

According to PULVER and VERZÁR (1941), horse leucocytes behave just like yeast cells with regard to the effect of glucose on the movements of K+. In their experiments, K+ entered the cells during the glycogen-building stage, and was subsequently released during glycolysis; no concomitant Na$^+$ movements were apparent, and erythrocytes from the same blood did not show the behavior at all.

Fragments of the hen's egg chorion were shown by Krogh (1943) to behave similarly to other tissues in losing K$^+$ in exchange for Na$^+$ when immersed in K$^+$-free Tyrode's solution, and in recovering upon addition of rather small concentrations of K$^+$. The K$^+$-deficient chorion took up K$^+$ from external levels as low as 1 mM, and expelled Na$^+$ against gradients as high as 150 mM. Following the pattern seen elsewhere, this activity required a reasonably elevated temperature, O$_2$, and a substrate such as glucose or lactate, although the rate of cation movements did not parallel the rate of substrate consumption.

Chambers et al. (1948) observed with K^{42} tracing that the K of eggs of two varieties of sea-urchins was largely not readily exchangeable prior to fertilization; but was almost entirely exchangeable at a fair rate, after fertilization. Low temperatures or cyanide reversibly depressed this exchange. None of these factors altered the total cell K content; however, in two Mediterranean urchins, Monroy Oddo and Esposito (1951) found a marked temporary uptake in total K within ten minutes after fertilization. Even when the only K$^+$ available in the medium was a small amount lost from the unfertilized eggs, a large part of it was briefly recovered shortly after fertilization. The nature of the mechanism involved remains unexplored.

Cation Concentration by Subcellular Particles

Recent work with subcellular particles *in vitro* suggests that the basis of the ion-accumulative properties exhibited by cells might at times reside in these particles rather than in a cell-surface transport system. The kidney-slice experiments discussed above are especially subject to such interpretation. Bartley and Davies (1952), analysing sheep kidney-cortex mitochondria after incubation at 20° C. with ATP, Mg^{++}, phosphate, and a suitable substrate, found such concentration ratios (particles/medium) as the following: H$^+$, 2.5; Na$^+$, 26; K$^+$, 2.0; Mg^{++}, 4.5; phosphates, 6.0; ATP, 0.7; pyruvate, 0.8; fumarate, 8.0; oxaloacetate, 0.1. They noted further (Bartley and Davies 1954) that these preparations maintained large K$^+$ and Na$^+$ concentration ratios for hours at 0° C. in a deficient medium (so that the *absolute* amounts were small); but that at 20° C. the extra Na$^+$ and K$^+$ were lost rapidly unless the ingredients mentioned above were available. Table IX illustrates some of these observations in a mixed particulate preparation. Bartley and Davies conclude from ATP^{32} experiments that ATP is the source of the mitochondrial P, the medium's orthophosphate not achieving direct entrance into the particle. The K^{42} exchange rate in this system even at 0° C. was too rapid to be estimated. The Na^{24} turnover at 0° C. indicated that about 20% of the Na$^+$ was exchangeable only with some difficulty, unless substrates such as fumarate or ketoglutarate were available.

Table IX. *Maintenance of Ionic Content of Sheep Kidney-Cortex Particulates* in vitro. (Adapted from BARTLEY and DAVIES, 1954.)

Treatment and Additions	Concentration ratios (particles/medium)	
	Na+	K+
"Cyclophorase" (nuclei and mitochondria)		
Aerobic; no addition	1.07	1.17
pyruvate, 5 mM.	1.24	1.14
ATP, 1 mM	1.15	1.10
both pyruvate and ATP.	1.43	1.05
Anaerobic; no addition	1.23	1.16
pyruvate, 5 mM.	1.07	1.12
ATP, 1 mM	1.33	1.15
both pyruvate and ATP	1.61	1.26
Mitochondria		
Untreated; 0° C	1.38	1.34
Twice washed in 0.9% KCl; 0° C.	26.5[1]	0.84
α-ketogluterate, 5 mM	1.17	1.24
and incubated 15 min. at 20° C.		
aerobically	1.54	1.62
anaerobically	1.41	1.37

[1] Na+ is here removed from medium, while almost none is lost from particles.

"Cyclophorase" preparations all incubated 30 minutes at 20°C. with Mg++ at 1 mM, NaHCO$_3$ at 12.5 mM.

STANBURY and MUDGE (1953) attempted to analyse the relation of the mitochondrial K+ maintenance to the metabolic activity in similar preparations from rabbit liver. Experiments with K^{42} showed a rapid, nearly complete exchange at 25° C., which failed at 2° C.; upsets in pH or osmotic pressure had to be rather extreme to affect the K+ accumulation, but ultimately caused a depression which paralleled the drop in the oxygen consumption. Ca++ in the absence of Mg++ also depressed both processes. However, orthophosphate excess depressed the K+ accumulation while enhancing the respiration. The use of DNP gave the most suggestive results, in that the maximal depression of the K+ content was at the concentration of DNP corresponding to the minimal Q$_{O_2}$ and P-esterification (lower and higher levels of DNP elevating these rates). BARTLEY and DAVIES (1954) point out that the energy of oxidative phosphorylation involved in their experiments is insufficient to give concentration ratios beyond about 1.03, if this is a system of active accumulation with passive escape. They suggest therefore that this is an exchange-diffusion type of system, with a metabolic coupling as an added complication.

MACFARLANE and SPENCER (1953) showed with rat liver mitochondria that adenosine-5'-monophosphate could provide the high-energy phosphate requirements for the K+ and Na+ accumulation in the presence of L-glutamate.

The report of Bartley, Davies, and Krebs (1954) gives an excellent summary of the information available on these matters, and on the newer concepts of the cellular distribution of low-weight molecules generally, notably in regard to the question of intracellular tonicity in the mammalian body, taken up in another section of this chapter. Mudge (1954) provides a review especially helpful in pointing out the immediate issues requiring further experimentation.

Transport of Inorganic Anions

Until the 1940's, it was generally believed that except for peculiarly anion-permeable cells like the mammalian erythrocytes, animal cell membranes did not allow passage of the negatively charged ions. In fact, the apparent impermeability to anions and to Na^+ provided the basis of classical bioelectric theory, by which the polarization of the cell membrane of muscle and nerve is attributed to the electrochemical equilibration of the diffusible K^+. Boyle and Conway (1941), finding the membrane permeable to Cl^-, interpreted the nearly complete exclusion of this ion under normal circumstances to the high concentration of organic anions which cannot escape through the cell surface, the Cl^- distribution then conforming to Donnan equilibrium characteristics. Except for the cells of organs especially concerned with transport functions, there has in any case been little reason to suppose that special metabolic assistance in passing through cell surfaces is given the ordinary inorganic anions of biological media, with the notable exception of the phosphates. Thus the primary subject of the first sections to follow will be the transfer through cell membranes of orthophosphate.

Phosphate Transport through Mammalian Red Cell Surface

In contrast to the freely penetrating smaller inorganic anions, orthophosphate seems to pass through the mammalian erythrocyte surface almost entirely in association with metabolic activity. In all this work, as stressed by Parpart and Hoffman (1954), it is essential to keep in mind the distinction between the true penetration process and the subsequent fixation of the phosphate into esters, such as occurs with active glycolysis. The latter involves actual total gain in "trapped" P, and should not be taken (as it sometimes has been) as indicative of an "active uptake."

By simple chemical analysis, Halpern (1936) showed a marked temperature dependence in the movements of inorganic phosphate into the human red cell from an external excess, and in the exit of phosphate into hypertonic solutions; these movements proceeded at an appreciable rate at 37° C., but failed at 3° C.[20] The addition of glucose enhanced the uptake of phosphate, which was given up again to the medium upon completion of

---

[20] Eisenman et al. (1940) obtained similar results in regard to the exchange of labelled phosphate, and suggested the transport had an enzymatic basis.

glycolysis, or when fluoride was added, the pattern being reminiscent of the behavior of K+ in many glycolysing cells as described in the previous section. HALPERN showed that the exit of phosphate from the cell may actually proceed against a concentration gradient. HEINSEN (1948) verified that the removal into the cells of extra phosphate added to human blood was associated with a disappearance of glucose, this relation not being observed with salts other than phosphates.

GOURLEY and GEMMILL (1950) studied this uptake with P^{32} and found that it followed first-order kinetics, at least over the first hour or so. The

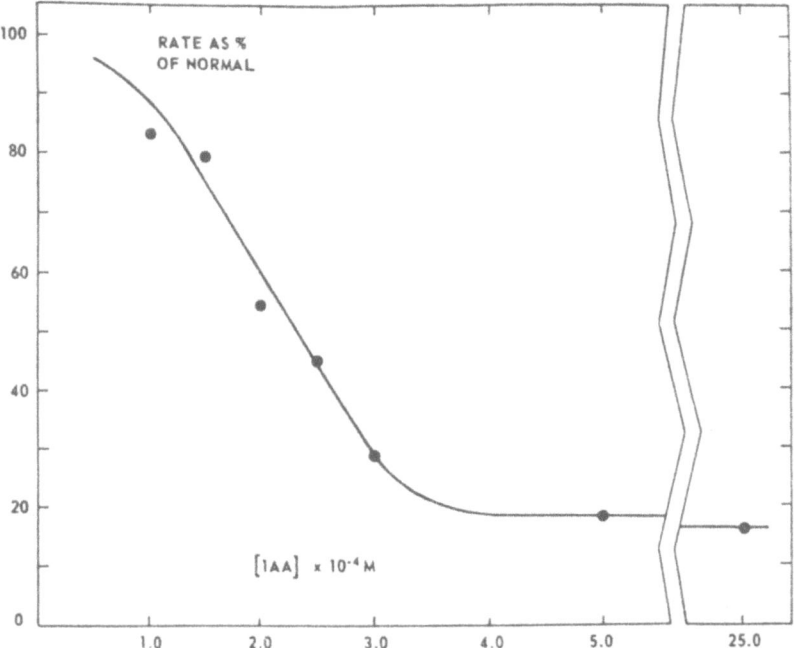

Fig. 14. Depression by iodoacetate of P^{32} uptake by erythrocytes. Inhibitor concentrations are final blood levels. (Courtesy of D. R. H. GOURLEY and American Journal of Physiology)

temperature dependency over the range 15–40° C. followed the Arrhenius equation closely, with $\mu = 16{,}700$ calories/mol, perhaps implying a chemical reaction; but anoxia had no significant effect on the rate of the process. HAHN and HEVESY (1942) reported the transfer was cyanide-sensitive, but not responsive to fluoride (see also MUELLER and HASTINGS 1951): GOURLEY (1951), on the other hand, showed that both iodoacetate and fluoride were effective inhibitors, although a residual process at about 20% of the normal rate could not be thus suppressed (see Fig. 14). The use of excess fluoride enabled GOURLEY to show that the uptake involves both a rapid, temperature-insensitive diffusion and a slower active process. Azide, Hg++, and DNP all failed to disturb the transfer. According to PERTZOFF and GEMMILL (1949), however, it is repressed by barbital or ether (not by some other anesthetics).

On the basis of the temporal course of the specific activity changes in the chief P-compounds of human red cells exposed *in vitro* to P[32]-labelled orthophosphate (Fig. 15), Gourley (1952) implicated ATP formation on the membrane as a step in entry of at least a significant fraction of the phosphate. The labile P of ATP and of diphosphoglyceric acid showed the most notable early rise of specific activity; on the same basis, in contrast to the situation in muscle as discussed in the next section, glucose phosphate did not appear to act as carrier for phosphate transport. Prankerd

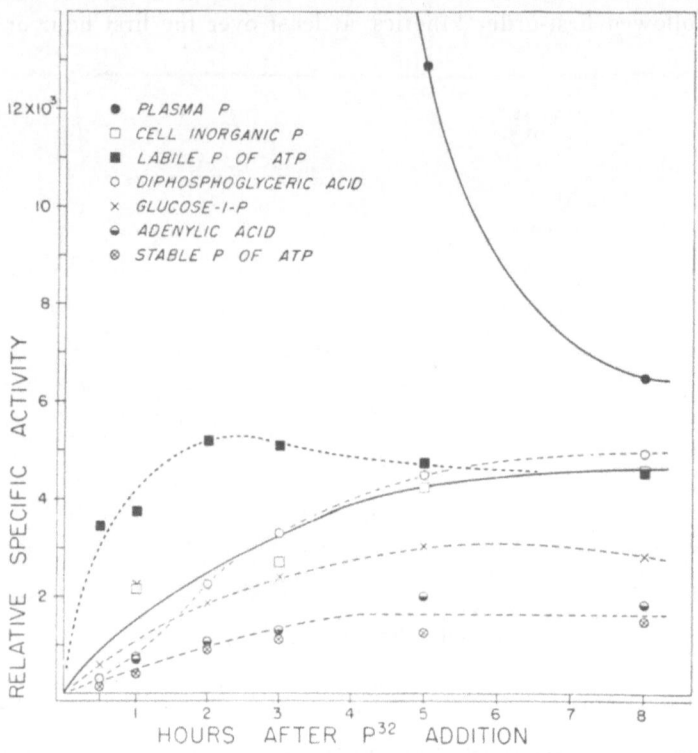

Fig. 15. Time course of relative specific activity of plasma inorganic P and the various P-compounds of human red cells. R. S. A. is given in counts/min./mg. P, recalculated to standard injection of 10[5] counts/min. per 100 ml. of blood.
(Courtesy of D. R. H. Gourley and Archives of Biochemistry)

and Altman (1954), using paper chromatographic methods to separate and identify the various organic phosphates, threw doubt on Gourley's finding of a special precursor role for ATP. Their data point clearly to the 2, 3-diphosphoglycerate as the component quickest to take up the P[32] label. However, recent unpublished observations with resin column separation techniques seem to support Gourley's original cruder experiments, and the interpretation of the discordant evidence has not been clearly settled.

On the basis of phosphate[32] uptake studies with rabbit red cells in a variety of mixtures *in vitro*, Jonas (1954) and Jonas and Gourley (1954) concluded that the process involves adsorption of the phosphate on a surface layer in a complex with Ca, protein, and lecithin, from which a reversible enzymic complex formation involving ATP, Mg[++], and K[+] carries the phosphate into the cell interior. ATP and Mg^{++} interacted in this system in such a way that when present in approximately equal concentrations they depressed the phosphate uptake, while a relative excess or deficiency of ATP enhanced the uptake. JONAS's arguments are developed

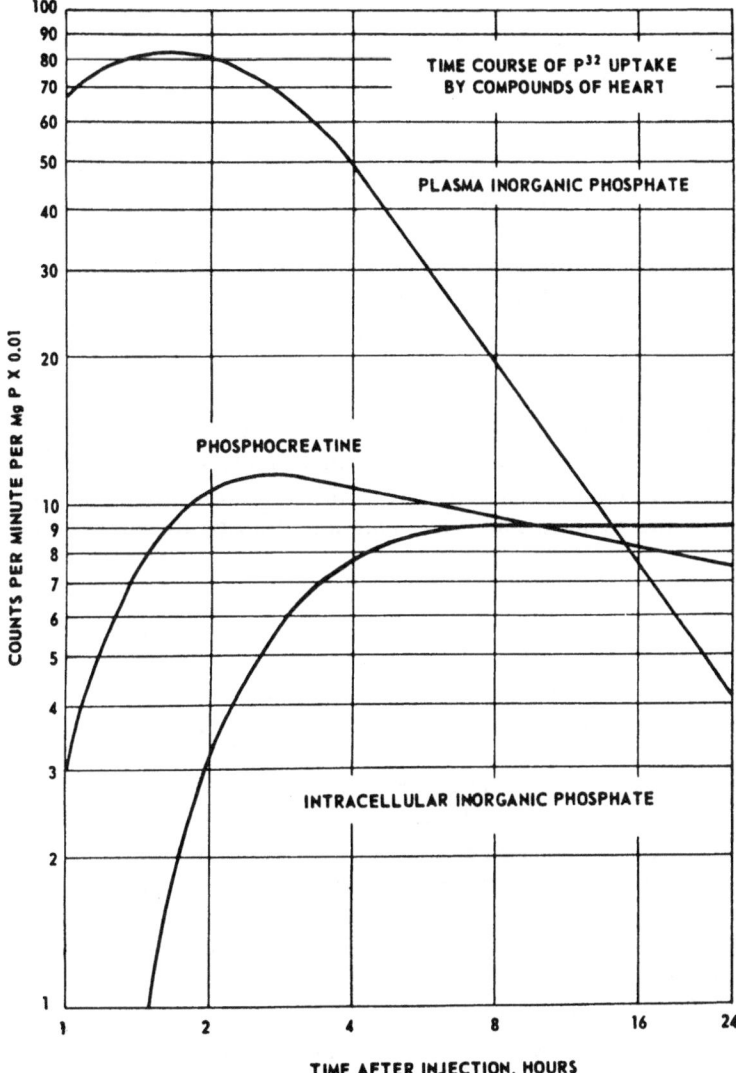

Fig. 16. Time course of P^{32} uptake by phosphocreatine and intracellular inorganic P of heart muscle of cats. Counts/min./mg. P are adjusted to standard initial injection 5 · 10^5 counts/min. per kg. of body weight. (Courtesy of J. SACKS and Cold Spring Harbor Symposia)

largely on the basis simply of the known physical and chemical make-up of the erythrocyte surface and consideration of the *possible* interplay of the ingredients by reason of their general chemical properties. Evidence of a more direct nature will of course be needed to establish these suggestions as a sound interpretation of the mechanics of the phosphate uptake.

Phosphate Transport through the Sarcolemma

The course of the early specific activity changes in the various phosphorus compartments, following the injection of $Na_2HP^{32}O_4$ into frogs or cats, was studied extensively by Sacks and Altschuler (1942), in an effort to reveal the interconversions of the organic P-bearing compounds. Of note to the question of anion transport was the fact that the early specific activities of the inorganic P of the tissues were consistently less than in the chief organic phosphates (see pattern in Fig. 16); this implies that the phosphate enters the cells by formation of organic compounds at the point of entry. Later (Sacks 1944 a, b) it was noted that in the fasting cat the specific activity of the muscle glucose-6-monophosphate (G-6-P) was higher at 2–4 hours after the P^{32} injection than at 24 hours, whereas that of the other primary organic phosphates continued to rise in relation to the G-6-P. This was taken to indicate that at least part of the G-6-P remained in ready contact with the extracellular compartment, and was probably held at the cell surface, as by adsorption. This "mobilization on the muscle cell membrane" of G-6-P was distinctly enhanced by glucose administration, and was seen only in fasted cats, not in "post-absorptive" cats. In view of the general evidence that glucose cannot freely diffuse into the muscle cells, these observations suggested that the formation of membrane-G-6-P is a step in the penetration (the sugar and phosphate entries thus being linked).

Administration of insulin at 5 U/kg. favored formation of the membrane-G-6-P, when carbohydrate reserves were depleted by fasting and muscular exercise (Sacks 1945), although no effect was apparent in resting animals. The turnover in a cat's heart (Fig. 16) is about twenty times as rapid as in its skeletal muscle, possibly by reason of this very response to activity (Sacks 1948). It appears from Furchgott and Shorr's (1943) report that this entrance of phosphate into the cardiac muscle by way of formation of organic phosphates is not absolutely essential to its penetration. In dog ventricular slices in labelled-phosphate Ringer, in which phosphorylation was inhibited by cold, no prominent organic P-compound was found to attain a given specific activity prior to the intracellular orthophosphate. Whether this was true at ordinary temperatures could not be ascertained by these workers because of the extremely rapid rates involved. (In liver, Sacks found still more rapid P^{32} exchange.)

Causey and Harris (1951) have provided evidence of an independent type in support of Sacks's principal thesis, showing by radioautography that a large part of the P^{32} taken up by frog sartorii is localized at the fibre surfaces.

Phosphate Uptake in Marine Eggs

In his early studies with P^{32}-orthophosphate, Brooks (1943 a, b) found in *Arbacia, Asterias,* and *Fundulus* eggs an irregular uptake involving a series of gains and losses, but achieving internal concentrations of as much as 6 times the level in the medium. The early embryos of the urchin and starfish behaved similarly to the eggs. ABELSON (1947) was able to achieve a much more stable and interpretable behavior with *Arbacia* eggs; be noted that, some 10 minutes after fertilization, P^{32} uptake increased by about 40 times, most of the P going into the acid-soluble fraction. The temperature dependency of this process was comparable to that of the cleavage rates. The uptake was cut to about $1/6$ by 4,6-dinitro-o-cresol at $1.6 . 10^{-5}$ M a concentration which doubles the respiratory rate and inhibits cell division. VILLEE *et al.* (1949) confirmed the bulk of these observations in *Arbacia* and looked into the action thereon of various other inhibitors, notably DNP and UO_2^{++}; the evidence was asserted to show that these agents acted on the uptake process itself rather than on any subsequent chemical conversion.

In the Pacific sea-urchins, *Strongylocentrotus purpuratus* and *S. franciscanus*, CHAMBERS and WHITE (1949) found much the same behavior, but noted also an actual net loss of P to the medium prior to fertilization. Most of the P^{32} acquired after fertilization was in the easily hydrolysed fraction containing the ATP, in which the increase more than accounted for the concurrent drop in the inorganic phosphate. LINDBERG (1950) extended this observation to the dejellied eggs of the Mediterranean urchin *Paracentrotus*, calculating that a layer of only 0.02–0.05 microns was involved in the free exchange which was self-terminating in about 30 minutes. The continuous accumulation into ATP ensuing upon fertilization involved rather rigid fixation of the phosphorus so that tracer was not lost when the source was removed from the medium.

BROOKS and CHAMBERS (1954) reported no such limitation in the uptake into unfertilized *Strongylocentrotus* eggs, although the rate was very low; a latency of 15–20 minutes preceded the large rate increase with fertilization. In *Urechis* eggs, the latency extends through the second cleavage, the rate then increasing through the next two cleavages. With external levels above 20 μg. of P per liter, the intake in *Strongylocentrotus* was not concentration-dependent, so that saturation of some other limiting factor is apparent. CHAMBERS and WHITE (1954) noted that in the fertilized eggs the flux calculated from P^{32} uptake equalled the net uptake, so that the efflux must be negligible. Here again it is likely that chemical fixation of the phosphate is obscuring study of the actual transport process.

Anion Transport in Microörganisms

Fluoride.—Changes in the "permeability" of microörganisms to fluoride ions have sometimes been suggested in interpreting observed changes in the inhibitory effects of this anion. On such a basis, RUNNSTRÖM and SPERBER (1938) and RUNNSTRÖM (1939) concluded that in baker's yeast the permeability to F^- was increased by anaerobiosis and by deficiency in the glucose supply. Brewer's yeast was more permeable in the first place, and was less subject to further change in this direction as a result of metabolic depression. However, on the basis of direct analyses MALM (1940, 1948) rejected this method except for the most crude qualitative comparisons, and concluded that the effects of glucose did not involve changes in the actual *uptake* of fluoride. Malm also noted that the Q_{10} of the actual uptake denoted a diffusion process.

Nevertheless, Runnström's line of argument was resurrected by Aubel and Sjulmajster (1950): they found that *E. coli* respiration was inhibited by F⁻ in the presence of succinate, and interpreted this to mean that the succinate permitted the F⁻ to enter the cell. In the absence of glucose, fluoride plus either malonate or azide also proved to be effective inhibitory combinations; but neither of the latter two agents assisted F⁻ in inhibition of glucose oxidation.

Phosphate.—Mullins (1942) showed that P^{32}-phosphate uptake in yeast was closely associated with carbohydrate metabolism. With higher temperatures and ample sugar supply, uptake was rapid and was little affected by anoxia. Like K⁺, the phosphate was later released to the medium as the metabolic activity subsided. Similarly, Caldwell and Hinshelwood (1951) found that *B. lactis aerogenes* on a glucose substrate removes inorganic phosphate from the medium prior to growth (so that the P content per cell increases), and releases the P later when growth ceases.

Spiegelman, Kamen, and Sussman (1948) observed that azide, without effect on glycolysis, essentially abolished P^{32}-phosphate anaerobic uptake by yeast, and slowed the disappearance of internal orthophosphate in phosphate-free media, without slowing its leakage into the medium. Thus the effect in this situation was not on "permeability" itself, but on the metabolic treatment of P, only secondarily affecting the uptake.

Riboflavin (at 1 mg. per liter) had a decided enhancing effect on P^{32} incorporation into two varieties of yeast (Nickerson 1948; Nickerson and Mullins 1948). In glycine-glucose or glycine-sucrose media, either aerobically or anaerobically, the vitamin increased the uptake by as much as 200%. The special significance of this observation lies in the fact that the intracellular content of riboflavine is much higher than this level, so that the effect would seem to involve some cell-surface process. Moreover, glucose assimilation and phosphate polymerization were also increased by the addition of riboflavine, so that a linkage of these processes to the P-transfer system (without involving respiration) is suspected. Nickerson (1949) showed a block by azide or DNP of all these linked processes, which could not be alleviated by riboflavine (Table X). The azide action has been confirmed by Conway and Moore (1950).

Uptake and fixation of orthophosphate by yeast entails concomitant uptake of cations rather than exchange for other anions (Schmidt, Hecht, and Thanhauser 1949). In these experiments, if a complete medium was presented, a strong preferential uptake of K⁺ was apparent; but, in the absence of Mg⁺⁺, Na⁺ was distinctly preferred. The presence of K⁺ augmented the rate of phosphate incorporation.

Roberts and Roberts (1950) found similarly that supplying K⁺ to K⁺-depleted *E. coli* greatly enhanced their incorporation of P^{32}. Analysis showed a fairly linear relation between cell [K⁺] and P^{32} incorporation, if other limiting conditions were made non-critical. However, the action of the intracellular K^+ level is probably not so much on phosphate-transfer processes as on synthetic activities; the incorporated P is not readily lost again, more than a mere translocation being involved. Glucose, fructose, or galactose assist in the phosphate uptake in this system, but G-1-P, G-6-P, or F-6-P are not effective.

Table X. *Phosphate Uptake and Metabolic Activity in Brewer's Yeast.*
(Modified from NICKERSON, 1949.)

Additions to glucose-phosphate medium	P^{32} Uptake (arbitrary units)	Glucose left unconsumed (mg./ml.)	Dry wt. change from assimilation (mg.)
None	247	13.8	63
DNP, $5 \cdot 10^{-4}$ M. . . .	0	10.6	— 5
NaN$_3$, $5 \cdot 10^{-4}$ M . . .	4.8	12.3	— 7
Riboflavin, $2.5 \cdot 10^{-6}$ M	314	17.0	70
Riboflavin and DNP . .	10.2	6.0	— 2
Riboflavin and NaN$_3$ . .	7.2	13.1	— 4

Figures are for 4 hours incubation, starting with 47 mg. dry weight of yeast and glucose at 106 mg./ml.

KAMEN and SPIEGELMAN (1948) noted the following general characteristics of the orthophosphate exchange in yeast and other microörganisms studied: (1) a high temperature coefficient; (2) sensitivity to azide, iodoacetate, arsenate; (3) less obvious sensitivity to fluoride; (4) correlation with cell metabolic activity; and (5) acceleration by previous exposure of the cells to a carbohydrate. Thus a considerable argument for a metabolically dependent transport system can be offered.

HOTCHKISS (1944) observed that *Staphylococcus* cells would not admit orthophosphate even with a high favorable gradient unless glucose was available; other well utilized substrates were not effective. In the presence of glucose, the uptake can proceed even to the point of building up an outward gradient. That the system affected is an actual surface transport is further suggested by the fact that the uptake is depressed by the antibiotic gramicidin, which is believed to be unable to penetrate to the interior of the cells.

Reabsorption of Anions in the Kidney Tubules

Phosphate.—WALKER and HUDSON (1937 b) showed by direct microsampling of the tubular fluid in *Necturus* and frog kidneys that certain individuals unquestionably reabsorbed phosphate actively in the proximal tubules, leaving the concentrations in the urine less than in the plasma; this was not, however, the *usual* situation. The frog's urinary output of phosphorus was studied in terms of the altered content of perfusion fluid in the glomerular and portal circulations by HOGBEN and BOLLMAN (1951).

They noted that the relation of phosphate transport to the load presented resembled an adsorption isotherm, and suggested that the "carrier" might be the tubular membrane itself. In dogs, Pitts and Alexander (1944) found the dependence of phosphate clearance on plasma levels was that of a typical threshold substance (see Fig. 17), with the maximal reabsorption on the order of 4 mg. P per minute. The avidity of the carrier system is

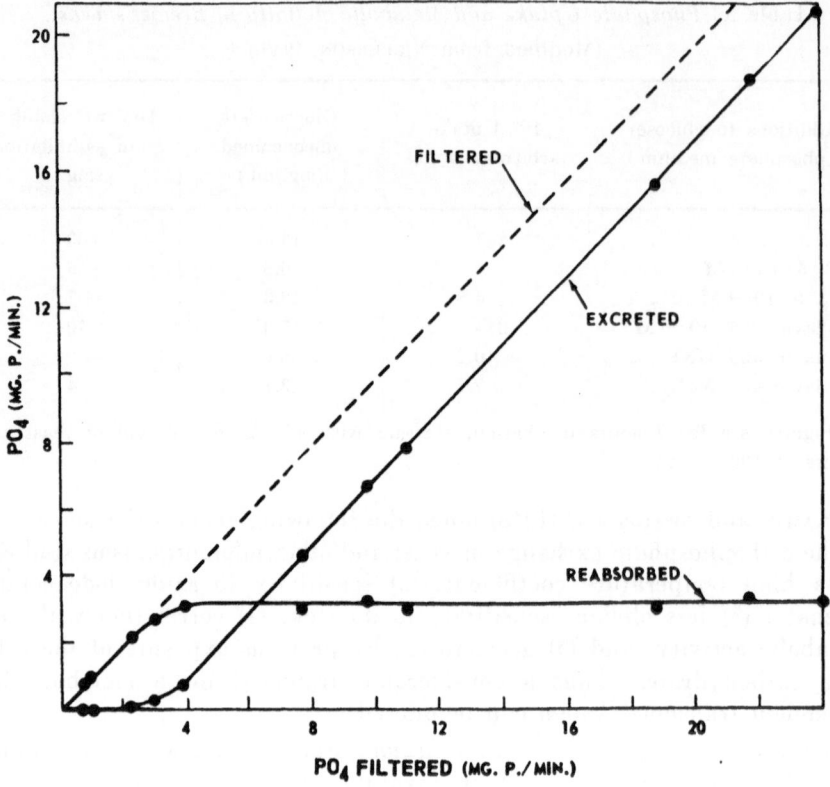

Fig. 17. The renal handling of phosphate in the dog, as a function of the load presented to the tubules in the glomerular filtrate.
(Courtesy of R. F. Pitts and American Journal of Physiology)

apparently not very high, since, as the figure shows, maximal reabsorption requires a plasma phosphate level significantly higher than the threshold concentration [21].

The significance of phlorrhizination to the phosphate reabsorption system has been a matter of some debate. Allan, Dickson, and Markowitz (1924) reported that in dogs phlorrhizin might cause a doubling of the urinary loss of phosphorus (following a transient slight drop). Lambrechts

[21] But Eggleton and Habib (1950) could find no clear renal threshold for phosphate in the cat, nor any limit to tubular absorption, although the process was depressed by the presence of PAH. Moreover, Eggleton and Shuster (1954 a, b) showed the rate of the process in the cat was improved by raising the plasma glucose level or by injection of insulin.

(1936 a, b) found, however, a nearly complete inhibition of the dog's phosphate excretion following phlorrhizin injection, even if measures were taken which ordinarily increase the P output. LUNDSGAARD (1935) did not observe any phlorrhizin effect on phosphate output in the lung-kidney-pump preparation. PITTS and ALEXANDER (1944) found that phosphate reabsorption was depressed by simultaneous reabsorption of glucose, although the phosphate system was itself unresponsive to phlorrhizin.

Chloride and Bicarbonate.—With the direct tubular fluid sampling procedure, WALKER et al. (1937) showed that the site of Cl^- reabsorption in the frog and *Necturus* kidney was the distal tubule. MONTGOMERY and PIERCE (1937) found in similar experiments that the acidification occurred also in the distal segment, and that the capacity of this system far exceeded any likely physiological demand. Bicarbonate absorption, rather than an actual H^+ or acid secretion, was felt to be the basic process involved. LOTSPEICH, SWAN, and PITTS (1947) noted that the overall rate of Cl^- and HCO_3^- reabsorption apparently has both active and passive components, and varies in parallel with the filtration rate, showing no definite maximum as does the phosphate transport process. The system has also been studied in humans (PITTS, AYER and SCHIESS 1948).

According to SCHWARTZ and WALLACE (1951), the diuretic action of mercurials in humans is probably due mainly to depression of tubular Cl^- reabsorption; patients on a mercurial regime often lost Cl^- in considerable excess of Na^+ and without change in serum $[Na^+]$. RICE et al. (1953) infused $NaNO_3$ into dogs and found no effect of mercuhydrin on the NO_3^- reabsorption, the action on the Cl^- system being fairly specific.

Sulfates.—The threshold for sulfate excretion in the dog is very sharp (LOTSPEICH 1947), and the clearance at high plasma levels approaches that of creatinine, so that apparently there is little or no diffusion component in the reabsorption. But no competition with chloride or phosphate transport was observed.

Thiosulfate, like the potassium ion, is apparently subject to both reabsorptive and excretory transport by the kidney tubule cells, although this would ordinarily be obscured by the fact that its clearance closely approximates the glomerular filtration rate. FOULKS et al. (1952) showed that under the influence of cortisone or testosterone, dogs excrete a large additional amount of thiosulfate, and that the drug carinamide (which depresses secretion of PAH, PSP and other *organic* anions) also reduces thiosulfate excretion to the point that its reabsorption becomes apparent. But a number of other inhibitors do not affect the system, nor does simultaneous handling of PAH. Thus the two sets of anion transport systems are evidently not identical.

Intestinal Absorption of Ions

It was established many years ago that salt solutions in the mammalian intestine are rather rapidly brought to a total tonicity not far from that of the blood; water and the chief inorganic ions migrate in the appropriate direction, and the bulk of the absorption then proceeds from a nearly isotonic solution. However, it was often noted that salt absorption does occur while the intestinal contents are hypotonic, and that water absorption occurs even from a hypertonic solution, so that uphill transfers were indicated. Clear demonstration of selective ionic transport against concentration gradients across the gut wall was provided by INGRAHAM and VISSCHER (1936 a), who followed the uptake from an equiosmolar mixture of NaCl and Na_2SO_4 in the tied-off lower section of the gut, in cats and dogs. Water and NaCl were removed from such mixtures, so that after about 1½ hours the Cl^- level might be as low as 0.5% of the blood $[Cl^-]$, although there was a rise in the luminal $[Na^+]$, $[SO_4^{--}]$, and total osmotic pressure. Such an extreme chloride depletion is of course far greater than would be accountable as a Donnan effect arising from the gut's relative impermeability to SO_4^{--}, and a genuine transport is indicated. However, it is not at all clear whether this involves directly the Cl^-, or the Na^+, or both. Use of K salts in place of the Na salts did not in essence alter the results. Moreover, if the polyvalent component in the mixture was a cation instead of an anion (INGRAHAM and VISSCHER 1938), the uni-univalent salt was still selectively absorbed so that in this case there was a marked depletion in the univalent cation. If the mixture contained equivalent amounts of both polyvalent cations and polyvalent anions, both species of univalent ions were nearly completely removed.

It was noted that this active uptake of univalent ions involved a considerable concurrent production of ammonia by the gut, but the connection between the two processes has not been elucidated. INGRAHAM and VISSCHER (1936 b) found that the specific uptake was abolished by application of a variety of metabolic poisons, including fluoride, hydrogen sulfide, Hg^{++}, and cyanide. Regardless of the direction of the net transfers occurring at any particular time, there is considerable reciprocal ionic exchange between the intestinal contents and the bloodstream, as shown by VISSCHER et al. (1944), with D_2O, Na^{24}, and Cl^{38}, in Thiry-Vella loops in the dog. The total gut Na^+ exchange can amount to about ¾ of the total plasma Na^+ per hour; the exchange in both directions is accelerated by elevated luminal $[Na^+]$.

Saturation kinetics in the absorption was demonstrated by BLICKENSTAFF et al. (1951), in that the absorption of Cl^- in the dog did not vary with the intestinal NaCl concentration over the range 85–217 mm. Again it is impossible to conclude whether the active process studied in terms of chloride actually involves this ion directly, or only secondarily moves the Cl^- in response to the transport of Na^+.

Anion Transport through Gill Epithelium

In analysing the urine and gastrointestinal fluids of a number of marine teleosts, SMITH (1910) crystallized the nature of the water and salt-regulating problem in these animals. He showed that the bulk of the salts other than $MgSO_4$ are absorbed from swallowed sea water, and that the kidneys excrete chiefly Ca⁺⁺, Mg⁺⁺, SO_4^{--}, and phosphates, leaving a hypertonic mixture of water, K⁺, Na⁺, and Cl⁻ to be eliminated by some other route. Later work has substantiated SMITH's tentative suggestion that the site of this excretory process is the gills. KEYS (1931), and BATEMAN and KEYS (1932) showed in an eel heart-gill preparation that Cl⁻ is extruded against a considerable concentration gradient into sea water; this extrusion decreases with the internal tonicity, so that the basis for a homeostatic mechanism is provided. The work done per unit mass of tissue in this process is on the order of that seen in the mammalian kidney, and may constitute a large part of the eel's resting metabolism. Cells of secretory type in teleost gills were described by KEYS and WILLMER (1932), but these were not found in the one elasmobranch studied.

Until the report of SCHLIEPER (1933), no clear consideration appears to have been given the question of whether the Cl⁻ is moved as a primary process or secondarily to a cation transport. SCHLIEPER tested the effects of adding Na_2SO_4, $NaNO_3$, and glucose to the inner and outer media in the eel heart-gill preparation, and concluded not only that the Cl⁻ was specifically extruded, but that its concentration in the internal milieu, rather than the total osmotic pressure, determined the activity of the system.

The gills may also serve in the *inward* transport of salts in a brackish or fresh-water environment. NAGEL (1934) came to the conclusion that this must occur in the shore crab *Carcinus maenas*, since although the antennal gland "kidney" continues to put out a fluid with the same osmotic pressure and chloride content as the blood while the tonicity of the environment is varied over a considerable range [22], the crabs are nevertheless fairly homoiosmotic. WEBB (1940) demonstrated by dialysis of the blood of this crab against sea water that the gills apparently maintain an internal excess of K⁺, Na⁺, Cl⁻, and Ca⁺⁺; the low blood Mg⁺⁺ and SO_4^{--} levels are accountable by the activity of the antennal glands.

KROGH (1937 b, c; 1938) showed that the gills may also carry chloride into salt-depleted crustacea, gastropods, mussels, and fresh-water fishes (Table XI), as does the skin in salt-depleted frogs. The eel failed to show this capacity. Since in the frog skin this behavior has proved interpretable on the basis of a demonstrated Na⁺-transport system, one naturally questions whether the gill activity studied in these other forms in terms of chloride movements is actually a chloride transport. When studied in terms of cations, as in MEYER's (1951) work with goldfish using Na²² as a tracer, the uphill transfer is equally apparent. KROGH observed that the bromide, thiocyanate, or cyanate salts were taken up actively as well as the chloride, whereas sulfate or nitrate could not be handled, and took this as an implication that the anion was not simply tagging along as an indifferent associate of an actively moved cation. But the parallel situation in the frog skin, where differential passive permeability to the anions can

---

[22] This is apparently characteristic of the antennal glands of the coastal crabs generally (JONES 1941).

alter the salt transport, highlights the weakness which Krogh recognized in this argument.

Table XI. *Absorption of Chloride into Salt-depleted Goldfish.*
(From Krogh, 1937 c.)

Duration in hrs.	Initial medium [Cl−] (mM)	Final Medium [Cl−] (mM)		Change rate, mg./hr.	
		Head side	Body side	Head side	Body side
5.75	1.36	0.713	1.750	− 0.24	0.26
5.2	1.16	0.11	1.22	− 0.43	0.05
1.0	0.954	0.739	1.020	− 0.46	0.25
1.0	0.954	0.747	1.026	− 0.44	0.29
3.0	0.984	0.518	1.125	− 0.33	0.19
3	0	0.0056	0.074	0.004	0.10

Rubber membrane around body separated head compartment (60 ml.) from body compartment (110 ml.).

Koch (1938) showed that following a washing in distilled water, *Culex* and *Chironomus* larvae take up salt from very dilute solutions, and this process was studied in terms of chloride ion absorption although no experimental demonstration was made that the anion was the actively transported element. By testing the effect of various ligations around the body, and localized chemical and thermal injuries, it was clearly established that the site of the active salt intake in this instance was the larvae's anal papillae ("gills").

In the terrestrial crab, *Ocypoda*, Flemister and Flemister (1951) demonstrated that the blood chloride level is successfully maintained at about 375 mM in the face of external variation over the range 120–600 mM. Urinary analysis showed that the antennal gland was either reabsorbing or excreting water as the occasion demanded, but that Cl− was lost at this site even into hypotonic environs. Thus by analogy with other aquatic animals and the lack of any other likely site, the gill was implicated as the organ for regulation of the chloride intake which must be responsible for maintaining the blood level. It was shown that the crabs' overall oxygen consumption rate was at a distinct minimum in a blood-isotonic medium.

Ronkin (1950 a) found that the uptake of P^{32}-orthophosphate into excised gill fragments of the clam *Mytilus* showed a turnover half-time of only about two hours, but that only a tiny fraction of the tissue P was involved in the exchange. As shown in Fig. 18, this activity responded to certain metabolic inhibitors, but not to others (Ronkin 1950 b). Since one of the effective depressing agents, iodoacetate, would be almost completely dissociated in the medium used, its action was believed to be strictly at the cell surface. This same argument is brought forward by Schoffeniels (1951) in a similar study on gill fragments of the mussel *Anodonta*; in this species, cyanide and fluoride were very effective inhibitors, bromoacetate

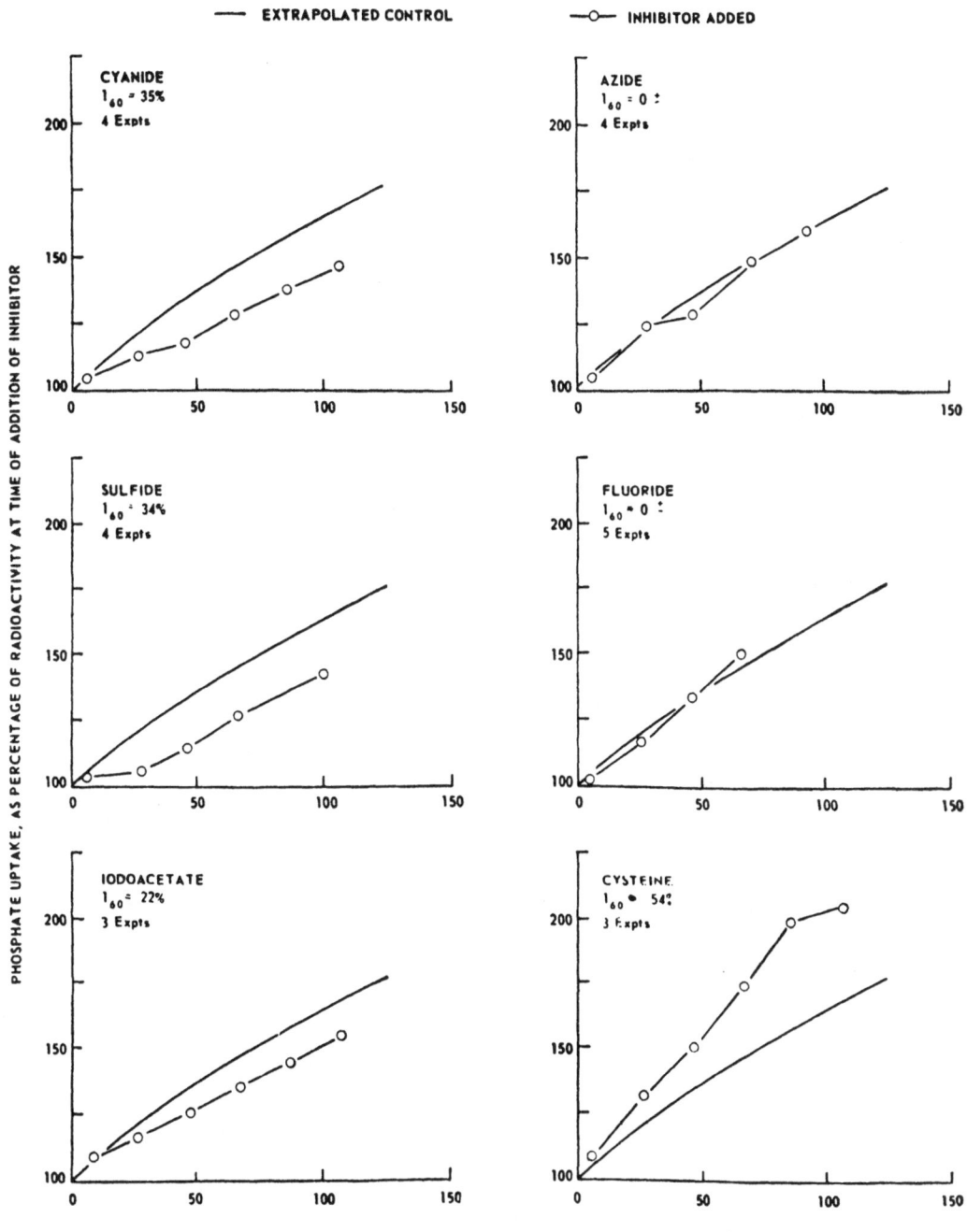

Fig. 18. P³² uptake in excised mussel gill as affected by various agents. Light line is expected result in absence of added reagent (average of 22 control experiments). l_{60} = % inhibition of uptake at 60-minute mark.

(Courtesy of R. R. Ronkin and Proceedings of the Society for Experimental Biology and Medicine) less so (all at 10^{-3} M, as in Ronkin's work).   Ronkin and Schoffeniels both noted that these transport inhibitors did not alter the ciliary activity of the gills.

Anion Transport in Other Cells

There has been little effort to analyse the uptake of phosphate into liver cells; Sacks (1951), in experiments with P^{32} of the type he carried out in studying muscle phosphate uptake, found that in liver the specific activity of the orthophosphate keeps somewhat ahead of that of the other components, so that no definite claim can be made for the role of phosphorylation in the process.   However, the labile P of ATP-ADP and G-1-P is not much slower in acquiring the label (all the rates are much higher than in muscle).   Popják (1950) reported that P^{32}-orthophosphate absorption into rabbit liver from an *in situ* perfusate was cut down by 50–75% by azide at 10^{-2} M, and concluded that the process did involve a membrane phosphorylation such as is generally accepted for muscle.

Grossfeld (1951), studying the swelling and shrinking of chick embryo cells in tissue culture in various sodium salts in relation to the lyotropic series of anions, observed that the volume changes were far slower under anaerobic conditions, which suggested that respiration might assist in the anion penetration.

Transport of Sugars and Related Compounds

Since animal cells do not typically have the capacity, common in the plant world, of synthesizing their fundamental organic materials from CO_2, NH_3, nitrates, and other inorganic molecules, it is apparent that they must take in, through their surface membranes, at least the simpler organic building blocks and foodstuffs needed to meet their energy requirements and provide for growth.   The three most obvious classes of such materials, for which a general need may be expected, are the monosaccharides, the amino acids, and the fatty acids.   The discussion to follow is accordingly taken up under these headings; incidental reference will occasionally be made to associated observations on the transport of other types of organic materials, but no attempt is made to present an inclusive summary of such information.

The classical empirical correlations of Overton, relating molecular physical properties to the ease with which the molecules can penetrate the typical cellular membrane, point to the sugars as relatively slow permeants. The monosaccharides are in general of a size which is just about the limit for any appreciable rate of penetration into most cells, and is apparently past this limit in many instances; furthermore, the sugars have a low lipoid solubility, and can be expected to be poorly penetrating on this count as well.   It is perhaps this conflict between the apparently fundamental structure of the plasma membrane and the necessity for penetration of the sugars that accounts for the evolution of special systems for handling the monosaccharides at the cell membrane; it was certainly the *awareness* of this conflict that led to much of the interest in looking for these systems.

Intestinal Absorption of Sugars

Kinetics of Absorption.—An early hint that the intestinal absorption of sugars might not be a simple passive process was given by the observation of NAGANO (1902) that the common monosaccharides were not all absorbed at equal rates from tied-off dog intestinal loops. CORI (1925), introducing what was to become a standard method permitting use of unanesthetized intact rats, further quantitated these specific differences in rates of absorption of the various monoses, finding the series:

galactose ⟩ glucose ⟩⟩ fructose ⟩⟩ mannose ⟩ xylose ⟩ arabinose.

In substance this sequence has been repeatedly confirmed, and it is now generally agreed that the first two (or perhaps three [23]) of these sugars are metabolically assisted in passing the gut wall. But there has been an unusual amount of disagreement concerning the dynamics of their absorption, possibly because so many different groups have taken up the study of the process. CORI claimed that the uptake from 50% solutions is steady at least until about 70% of an administered dose is removed from the lumen. He concluded that the concentration of sugar presented is not a significant factor in the rate of absorption (see also CORI, CORI, and GOLTZ 1928). This is of course one of the chief types of evidence for characterizing a translocation system as an active transport.

TRIMBLE, CAREY, and MADDOCK (1933) confirmed this finding. In extending the observations to lower concentration ranges, AUCHINLACHIE *et al.* (1930) found, as would be expected, a saturation concentration (about ¾ M) below which the glucose uptake did vary with the concentration. The significance of this was made more secure by the fact that no such saturation concentration was found for xylose, nor, after death of the tissue, for either sugar. As shown in Fig. 19, VERZÁR (1935) also found the saturation type of kinetics with galactose or glucose, and a direct concentration-dependence (as would be expected for passive diffusion) with mannose, sorbose, or xylose; the behavior of fructose was intermediate. But both MACKAY and BERGMAN (1933) and FEYDER and PIERCE (1935), using CORI's methods, disagreed as to the behavior of glucose, claiming that the uptake diminishes with time while the residual amount is still considerable, and that furthermore the initial rate increases appreciably with the concentration. FENTON (1945) also insisted that the glucose absorption in rats increases with the level of glucose (though not rectilinearly) all the way up to 65% solutions, and that the rate decreases with time under any given set of conditions. FENTON attacked the whole basis of the standard CORI procedure, showing that the factor most concerned in the apparent "ab-

---

[23] Fructose is apparently entirely converted to glucose in the course of its absorption, at least in the guinea-pig (DARLINGTON and QUASTEL 1953).

sorption rate" with this method is the gastric emptying time [24]. When glucose solutions were mixed with saline so as to maintain isosmoticity in all mixtures, it was found (Wix *et al.* 1952) that there was an optimal glucose concentration for its absorption in the rat, at about 3%.

GROEN (1937) tested many of the conclusions of the rat studies in the human by intubation of the gut and sealing off a given length with a balloon. The relative rates of uptake of galactose, glucose, and fructose proved to be as in the rat. The absorption was steady with time and was

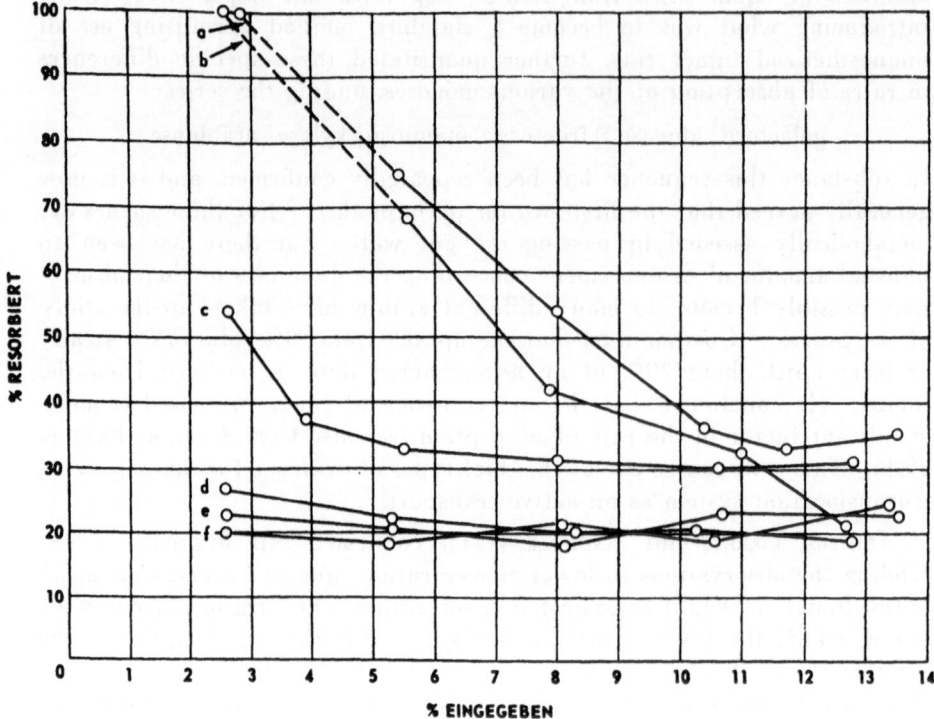

Fig. 19. Relation of percentile sugar absorption to luminal concentration. 3 cc. of solution in tied-off upper 30 cm. of rat gut *in situ*. Period of absorption was one hour. *a*, glucose; *b*, galactose; *c*, fructose; *d*, xylose; *e*, sorbose; *f*, mannose.
(Courtesy of Biochemische Zeitschrift; from VERZÁR, 1935)

independent of concentration above isotonicity (but the fact that uptake of water reduced all hypertonic concentrations to the isotonic level could account for this). Only extremes of pH had any effect, and starvation or various diets did not alter the rates. MINIBECK (1939) found a similar sequence for the sugars' relative absorption rates in frogs.

In studies with an isolated rat gut perfusion technique, FISHER and PARSONS (1953 a, b) showed that the glucose and galactose absorption followed a MICHAELIS-MENTEN pattern, the reciprocal of the rate bearing a rectilinear relation to the reciprocal of the luminal sugar concentration.

---

[24] LOURAU and LARTIGUE (1951) also present cogent criticisms of the "Cori coefficient" as applied to guinea-pigs.

On such a basis, they calculate the MICHAELIS half-saturation constant for the enzymatic carrier as 8 9 mM for glucose and about 35 mM for galactose, although the maximum transport velocities are not markedly different for the two sugars [25].

Competition between Sugars.—CORI (1926 a) noted apparent competition for the same carrier mechanism, in that the total absorption from mixtures of glucose and galactose was about the same as that from the pure solutions of each alone. However, in line with the apparent passivity of the xylose kinetics, MACLEOD, MAGEE, and PURVES (1930) showed that the uptake of glucose and xylose do not interfere with each other when occurring simultaneously.

In their rat-gut perfusion set-up, FISHER and PARSONS (1953 b) observed a much greater inhibition of galactose than of glucose uptake when both were present in equivalent amounts. However, on the basis of the relative MICHAELIS constants noted in the preceding section, the depression of galactose uptake should have been appreciably greater than it was. Taken together with the fact that the anatomical longitudinal gradient of absorption rates was not identical for the two sugars, this suggested that the common pathway did not account for all the absorption observed. This recalls the similar situation noted by LeFEVRE and DAVIES (1951) in regard to glucose and galactose uptake into human red cells.

WESTENBRINK (1934) made the curious observation that absorption of a given one of the actively transported sugars by rats was favored by prefeeding the same sugar. Therefore, he retested CORI's quantitative comparison of the absorption rates of the several sugars using rats on a carbohydrate-free diet, but found the same general arrangement here, as well as in the pigeon (WESTENBRINK 1937; WESTENBRINK and GRATAMA 1937). These investigators questioned the passivity of xylose absorption, since the pigeon absorbed D-xylose only about half as rapidly as L-xylose, which was taken in at a rate comparable to that of L-arabinose.

Association with General Metabolism.—HEWITT (1924) showed that when cat gut epithelium was killed, the selective uptake rate differences among the hexoses disappeared. AUCHINLACHIE, MACLEOD, and MAGER (1930) brought out significant differences in the absorption of xylose and glucose in isolated rabbit gut loops in Tyrode's solution. In living loops at 40" C., the glucose uptake is the faster, while if the temperature is lowered to 0" C., or if the loop is killed by heat, NaCN, or NaF, the xylose moves out more rapidly. VERZÁR and WIRZ (1937) established that the glucose uptake from rat gut loops is decidedly temperature-sensitive over the range 24 40" C., while the xylose uptake is not. When the active process was depressed by lowering the temperature to 27" C., the glucose uptake became nearly proportional to concentration (as was that of xylose even at body tempera-

---

[25] These authors stress the intrinsic meaninglessness of the classical comparisons of the absorption of two or more substances in terms of a mere rate ratio. when the concentrations used and the nature of the kinetics are not given.

ture), so that a passive component in the glucose movement is also likely. In line with this, it was noted that iodoacetate reduced the glucose-absorption temperature coefficient.

Sivilla (1953) showed that hypertonicity depresses absorption from gut loops of only the actively transported sugars. He noted that even the moderate hypertonicities as used by the Cori school constitute a serious disturbance, so that the true differences between the selective and non-selective rates are even greater than has been generally appreciated.

The Role of Phosphorylation.—Magee and Ried (1931) noted that addition of a small amount of phosphate accelerated the glucose uptake by about 50%. But in confirming this in connection with their suggestion that the sugar might be phosphorylated as a prerequisite to absorption, Wilbrandt and Laszt (1933) failed to find any change from the starvation level of hexose phosphate in the intestinal epithelium during absorption. Laszt (1935) found the phosphate effect maximal (increase of 65—100% in the rate) at pH 7.0, and a similar action with borate or acetate buffers; this effect fell off on either side of neutrality, being essentially absent at pH 4.0 or pH 7.8. Xylose absorption did not respond to these influences.

Laszt and Süllman (1935) analysed the P-fractions of the rat intestine and found a definite increase in the acid-soluble organic P during absorption of glucose, fructose, galactose, or glycerol, but not with mannose, xylose, or arabinose; sorbose was intermediate and doubtful. Lundsgaard (1939) reported a drop in inorganic P in rat intestinal mucosa during absorption, and a concurrent rise in readily hydrolysable organic P, these effects being greater with fructose than with glucose. In cats and rabbits, this whole phenomenon was absent or virtually so. Application of cyanide immediately blocked these reactions in the rat, so that three minutes later, the inorganic P levels were already restored, the organic P elevation being about half what it had been (Kjerulf-Jensen and Lundsgaard 1940: Table XII). These rates are sufficiently rapid that the phosphorylation idea is reasonable insofar as the temporal characteristics of the process are concerned. Beck (1942a) found the mucosal P-fixation with glucose absorption to be largely in the form of organic pyrophosphates. However, this fraction was still formed in the presence of phlorrhizin, while the increase in the difficultly hydrolysable fraction (including any hexose monophosphate esters) was largely abolished by phlorrhizination.

Nakazawa (1922) working with rabbit gut, was probably the first to show that phlorrhizin prevented absorption of glucose without affecting absorption of water, salts, fatty acids, and other molecules. This phlorrhizin effect was confirmed by many others in the early 1930's and was extended to the other common monosaccharides by Wertheimer (1933), who found that the inhibition in rats and mice applied to glucose, galactose, and possibly fructose, but not to the sugars with slower specific rates

Table XII. *Changes in Acid-Soluble Phosphate of Gut Mucosa During Sugar Absorption.*

(From KJERULF-JENSEN and LUNDSGAARD. 1940.)

Time	P concentration, in mg. per 100 g. dry wt. of mucosa				No. of expts.
	Inorganic	Easily hydrolysable	Difficultly hydrolysable	Total	
At rest.	131	141	196	460	21
During fructose					
uptake	74	238	192	518	15
Change.	— 57	97	— 4	58	
15–30 seconds after					
cyanide injection . .	104	213	204	520	10
Change.	30	— 25	12	2	
30–60 seconds after					
cyanide injection . .	112	198	196	501	5
Change . \	38	— 40	4	— 17	

of uptake [26]. He noted that absorption from the peritoneal cavity, which shows none of the specificity of the process in the intestine, was likewise insensitive to phlorrhizin. According to ABDERHALDEN and EFFKEMANN (1934), the action of phlorrhizin is duplicated by the other glucosides amygdalin, arbutin, and salicin. DONHOFFER (1935) found in the phlorrhizinized rabbit no residual glucose absorption whatever if the luminal sugar level was only about M/10.

WILBRANDT and LASZT (1933) observed in the rat that iodoacetate also markedly reduced glucose and galactose uptake and slightly inhibited the uptake of fructose or mannose, but not of the pentoses (Table XIII). In contrast to this, lethal doses of NaCN, urethane, or NaF had no *specific* effect on the hexose uptake rates, but lowered absorptive activity generally. In view of the *in vitro* inhibition by iodoacetate of glucose phosphorylations, this suggested that such a step was an essential part of the intestinal absorption process.

VERZÁR and LASZT (1935 a) showed that iodoacetate reduced the uptake of glucose without altering the simultaneous absorption of Na_2SO_4 or sorbose. But WESTENBRINK (1937) insisted that iodoacetate blocked xylose uptake as well as glucose uptake, and he asserted that the action is simply on the circulation. KLINGHOFFER (1938) agreed that iodoacetate depressed xylose absorption, even salt absorption; and that there was

---

[26] BOGDANOVE and BARKER (1950) claimed phlorrhizin inhibition was most notable with sorbose, far less with other hexoses, and absent (actually an acceleration) with fructose, but this is contrary to general experience.

severe gastrointestinal pathology (and death within 8 hours) when this agent was applied at the concentrations used by the Verzár school. Öhnell and Höber (1939) confirmed this, noting that the less drastic phlorrhizin effect was more indicative of a true transport inhibition.

Table XIII. *Depression of Sugar Absorption by Iodoacetate in Rats.*
(From Wilbrandt and Laszt, 1933.)

Sugar	Percentage Absorbed in 1 Hour		Ratio Unpoisoned/poisoned
	Untreated	With Iodoacetic Acid	
Galactose	84.3	39.6	2.1
Glucose	73.3	24.3	3.0
Fructose	32.5	27.5	1.2
Mannose	23.9	18.6	1.3
Xylose.	21.8	22.8	1.0
Arabinose	21.0	21.5	1.0

Averages of 4–6 runs; iodoacetic acid given subcutaneously at about 0.15 mg. per g. body weight.

Brückner (1951) stressed this confusion in interpretation of previous use of inhibitors in this field by reason of the general toxicity effects, which had been largely ignored. He found inhibition with DNP or atabrine, if at all, only at levels which induced non-specific cell damage. However, the case for specific inhibition by phlorrhizin seemed acceptable; with this agent at only 10^{-4}-10^{-3} M, he showed marked depression of galactose or glucose (but not sorbose or xylose) absorption from rat gut loops. On the basis of the actions of these three inhibitors on isolated enzyme systems, Brückner suggested that the critical step depressed by phlorrhizination is not the sugar phosphorylation itself, but the dehydrogenase activity essential to the building up of ATP.

Fridlander and Quastel (1953) reported that O_2 lack or application of DNP [27] depressed only the selective transport in the guinea-pig gut (glucose, galactose, and converted fructose, but not sorbose). Wix, Fekete, and Horváth (1951) noted no effect on rat glucose absorption, with DNP at 3–5 mg.%.

Mathieu (1935) had found the absorption of G-1-P, G-6-P, or G-1,6-diP in the rat to be on the order of that of fructose at the same concentrations. This was reinterpreted, however, by Rothstein, Meyer, and Scharff (1953); they showed that rat gut loops rapidly hydrolyse G-1-P, leaving orthophosphate in the lumen and reabsorbing glucose, the G-1-P itself being unabsorbable. G-1-P[32] tracing proved that the P did not leave the lumen; since phosphatase did not enter the lumen, the action must have been at the cell surfaces in contact with the luminal contents.

---

[27] Darlington and Quastel (1953) added cyanide, azide, chloretone, fluoroacetate, malonate, and quinine to this list.

The Reality of the "Pump."—BÁRÁNY and SPERBER (1939) used Na_2SO_4 or sorbose as osmotic fillers in dilute glucose solutions in the rabbit gut, and showed by direct analysis that terminally the glucose was being moved up a chemical gradient into the blood stream.

FISHER and PARSONS (1950, 1953 a) used a closed perfusion system on small lengths of rat intestine [28] and, beginning with equal concentrations (0.4%) inside and outside, showed a definite outward glucose pump, building up a gradient. This was blocked by phlorrhizin at 0.1%. Increasing the glucose level on the serosal side did not markedly depress the removal from the lumen, although a considerably larger part of the sugar taken up remained within the gut tissue itself.

LOURAU and LARTIGUE (1951) noted that in x-irradiated rats there was a rather sudden drop in absorption after experimental runs had been under way about a half-hour; this was correlated with the appearance of hyperglycemia under these circumstances, and it was suggested that the blood level was acting like the "trans"-concentration in WILBRANDT's scheme for the blood cell sugar transport system (discussed later), thus reducing further absorption. However, HESTRIN-LERNER and SHAPIRO (1953) found that glucose disappearing from rat gut lumen is not accounted for by its appearance in blood *in situ,* or in perfusate or external solution *in vitro.* C^{14}-glucose tracing showed a substantial fraction is converted to a non-reducing non-fermentable substance which does not release glucose upon hydrolysis.

WILSON and WISEMAN (1954) everted the small intestine of rats or hamsters to form collecting sacs for the study of the transference of glucose and other substances through the gut wall. The uphill movement of glucose into such sacs failed under anaerobic conditions. Since a large part of the metabolized glucose turned up as lactate on the serosal side, the question was raised whether this might not be HESTRIN-LERNER's unidentified substance.

Effects of Endocrine Factors.—A great deal of information is available concerning the relation of various hormonal factors to the selective absorption of sugars by the intestinal epithelial cells. Since little appears to be known about the means of action of the hormones at the site of absorption, it is not appropriate to go into these matters here. The adrenals and hypophysis seem unquestionably involved in the maintenance of the selective monosaccharide absorption extensively studied in the rat, and significant observations have been made on the influence of most of the other primary endocrine organs. For an especially thorough summary of all but the most recent items, the reader is referred to SOULAIRAC's (1947) treatise on the neuro-endocrine regulation of glucose absorption in the rat.

---

[28] This technique permits separation of the appreciable intestinal sugar *utilization* from the true *translocation.* HORVATH and WIX (1951) have developed also an *in vivo* luminal perfusion system for this type of study.

Renal Tubular Absorption of Sugars

The Reality of the Active Process.—Using the classical double perfusion technique which takes advantage of the separation of the glomerular and tubular circulations in the frog kidney, Clark (1922) showed a complete glomerular permeability to glucose even at subthreshold levels, and a reabsorption at the tubules even when the capillaries contained glucose at 9–10 times the normal blood level. Ca^{++} in the perfusate was necessary to maintain this uphill transport. Walker and Hudson (1937 a) directly followed the reabsorption of glucose by sampling and analysis of the fluid in the tubules of *Necturus* and frog kidneys. Beginning at the glomerular end of the tubule, the reducing power of the fluid fell rapidly until it became negligible before reaching the distal tubule. By perfusing the distal tubule with a glucose-Locke's solution it was shown that this segment lacked the ability to absorb glucose that was shown in the proximal tubule. Wood (1941) demonstrated by the same procedures that the reabsorption in *Necturus* was not depressed by considerable elevation of the plasma sugar level, even though the tubule received by perfusion a more dilute (previously collected) glomerular filtrate.

Gammeltoft and Kjerulf-Jensen (1943) reported that man has a greater capacity to reabsorb galactose and fructose at low plasma levels than have cat, dog, or rabbit, but that the tubular maximum in man is rather low. Cats apparently fail to show any clear tubular maximum for glucose (Eggleton and Shuster 1954 a); the urinary loss of some of the filtered load, seen at very high plasma levels, was thus attributed to a depressing effect of the high glucose levels rather than to actual saturation of transport processes. Injection of insulin, it may be noted, enhanced the already nearly complete reabsorption (Eggleton and Shuster 1954 b).

Competition between Sugars; the Mechanics of Reabsorption.—Shannon (1938) showed that xylose and glucose vie for the same reabsorption carrier mechanism in the dog, glucose being taken back preferentially. According to Gammeltoft and Kjerulf-Jensen (1943), loading with glucose also results in a depression of reabsorption of galactose or fructose. Thus the scope of the system differs considerably from that described for the intestinal absorption; it resembles more nearly that to be presented for the erythrocyte membrane in the next section.

Shannon and Fisher (1938) proposed that the glucose uptake involves a reversible reaction with a tubular cellular element of which there is a constant limited amount, and that the sugar-carrier complex breaks down at a rate so slow relative to the first reaction that the tubular sugar and the carrier are always essentially in equilibrium. This of course fits the "saturation kinetics" observed in the process (Fig. 20).

The Role of Phosphorylation.—The well-known action of the glucoside phlorrhizin in promoting glycosuria was interpreted many years ago in terms of an inhibition of the tubular reabsorption. Poulsson (1930) showed that following phlorrhizination of dogs, their urine/plasma concentration ratios for glucose and creatinine came into exact parallel, and the total glucose elimination rate did not vary appreciably even though the urine

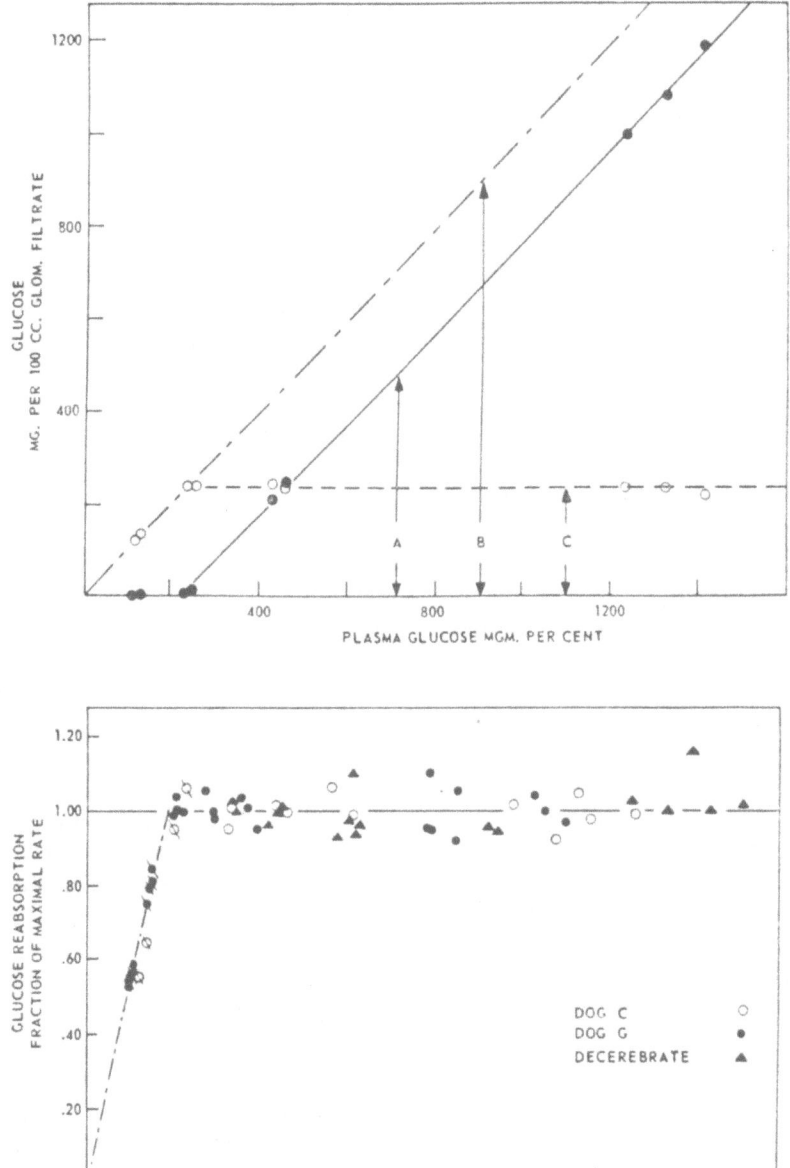

Fig. 20. Glucose transport in the dog kidney tubule. Above; A (solid dots) — rate of excretion in urine. B — rate of filtration, C (open circles) — rate of reabsorption, by difference between B and A. All referred to filtration rate as estimated from creatinine clearance. Below: Reabsorption as a function of plasma level in three dogs. All values are referred to mean of observations on same animal at plasma levels giving frank glycuresis.

(Courtesy of J. A. SHANNON and American Journal of Physiology) volume output might fluctuate greatly, so that clearly tubular activity on the glucose had been sharply curtailed.

LUNDSGAARD (1933 a) was the first to suggest that this action of phlorrhizin might have its basis in the inhibition of enzyme activities essential to the transfer process; he observed that phlorrhizin at 0.005–0.02 M was an active inhibitor of phosphorylation and dephosphorylation in various tissues. Upon injection of a diabetogenic dose in rabbits, such concentrations of phlorrhizin were found only in the kidney; this supported the suggestion that a phosphorylation might be involved in the normal transport process in the tubule wall, just as had been suggested by WIL-BRANDT and LASZT in the interpretation of their experiments with iodo-acetate effects on sugar absorption in the gut.

However, LAMBRECHTS (1934) found in the dog kidney that the total organ concentration of phlorrhizin at doses eliciting glycosuria did not attain phosphorylation-inhibiting levels. Moreover, in LUNDSGAARD's (1935) studies in the lung-pump-kidney preparation, he found a complete block of glucose transport at a phlorrhizin level (1 mg. per kg. of kidney) which would be effective against the most sensitive enzyme systems only if it were restricted to no more than $^1/_{10}$ of the kidney mass. LAMBRECHTS (1936 a) similarly concluded that either there must be such a localized accumulation of the phlorrhizin or else its action was not actually attributable to inhibition of phosphatases; since, in comparing a great number of phlorrhizin relatives, he could establish no parallel between their relative effectiveness as glycosurics and their activity on dog kidney-mash phosphatases, he was inclined toward the latter view. Also, he noted (LAMBRECHTS 1937) that phlorrhizin seriously hampers resorption of rhodamine, which cannot be phosphorylated at all. With their direct sampling method, WALKER and HUDSON (1937 a) noted that after phlor-rhizination, the reducing power of the fluid actually increased in passing down the tubule, as water was reabsorbed without equivalent glucose. This was observed at phlorrhizin levels considered too low to inhibit the phosphatases; moreover, well-phlorrhizinized rat kidneys showed the same phosphatase activities as normal kidneys.

On the other hand, BECK (1942 b) showed that rabbit kidney-brei glucose phosphorylase was somewhat more sensitive to phlorrhizin than were other similar enzymes. Also, by microscopic observation of frog and rat kidneys *in situ*, ELLINGER and LAMBRECHTS (1937) secured direct evidence of heavy concentration in the tubules of several colored azo-derivatives of phlorrhizin after intravenous injection. An inactive derivative, passing the glomerulus, was concentrated in the tubular lumen and not resorbed, whereas two active forms became only slightly concentrated in the lumen, collecting densely for several hours in the luminal border of the second segment of epithelial cells, which presumably includes the site of the active process.

Histochemical studies have also suggested the importance of phosphatases in the renal tubular function. KABAT and FURTH (1941) showed a concentration of phosphatase activity in the brush borders of the convoluted tubules; many other similar observations are cited by ROTHSTEIN in his monograph in this handbook. SOULAIRAC, DESCLAUX, and TEYSSEYRE (1949) found that this enzyme concentration gradually disappeared with the development of glycosuria after alloxan treatment, while insulin reversed these changes; and WILMER (1944) noted that the aglomerular toadfish's kidney tubule shows no phosphatases histochemically. However, KRITZLER and GUTMAN (1941) found no activity differences in the kidney tubule "alkaline" phosphatase in rats acutely or chronically phlorrhizinized, nor in biopsy material from dog kidneys phlorrhizinized locally *in situ*, although there was frank glycuresis [29].

Sugar Transport in the Human Red Cell Surface

The Kinetic Peculiarities.—It was established in the early permeability studies with mammalian erythrocytes that apparently only the primate red cells allowed the monosaccharides to penetrate at appreciable rates. The first indication that this process did not obey simple diffusion laws came evidently with EGE's (1919) thesis at Copenhagen, cited by BANG and ØRSKOV (1937); EGE showed that the rate of uptake of glucose into human red cells did not increase with its concentration in the medium to the extent predicted. But this was attributed mainly to a probable loss of K^+ such as has been noted above as a common response to the addition of non-electrolytes.

There has never been any report that sugars can be actually *accumulated* by red cells against a concentration gradient. KOZAWA (1914) and EGE and HANSEN (1927) showed that the blood glucose in man is evenly distributed in the water of the plasma and corpuscles. Moreover EGE, GOTTLIEB, and RAKESTRAW (1925) reported rapid equilibration of small amounts of added glucose, and this made it difficult to understand why the red cells do not hemolyse osmotically in pure isosmotic glucose solutions. KLINGHOFFER (1935) more clearly defined the nature of this paradox by showing that the ready uptake occurred only in concentrations of glucose below about 2 percent.

BANG and ØRSKOV (1937) found that the "permeability constant" in this system varied inversely with the glucose concentration, over a small range. GUENSBERG (1947) greatly extended this observation to a range of glucose concentrations of over a hundredfold [30]. He believed the depression of the proportionate uptake at the high sugar concentrations might derive from an action of the sugar on the membrane's permeability,

---

[29] According to MARSH and DRABKIN (1947), phlorrhizin at 10^{-2} M inhibited rat kidney "alkaline" phosphatase far more than the "acid" phosphatase. Addition of extra phosphatase to kidney homogenates cut down on the phosphorylation, and since alimentary hyperglycemia markedly raised the activity of both phosphatases, it was suggested that this might constitute a mechanism for lowering the glucose threshold under these circumstances.

[30] A less complete but more widely available note on this point is that of WILBRANDT, GUENSBERG, and LAUENER (1947).

rather than from operation of a carrier system, such as suggested by LeFevre (1948) in interpretation of his parallel observations.

LeFevre and Davies (1951) showed that the aldoses D-glucose, D-galactose, D-mannose, D-xylose, and L-arabinose all exhibit this "saturation kinetics," whereas the ketoses L-sorbose and D-fructose appear on this basis to move passively. Fig. 21 gives examples of each type of behavior. These differences are the source of the discrepancy between Kozawa's (1914) and Wilbrandt's (1938) sequences of entry rates for these same sugars, since the two were working with very different sugar concentrations. For reasons developed below, it appears that both the ketoses and aldoses share

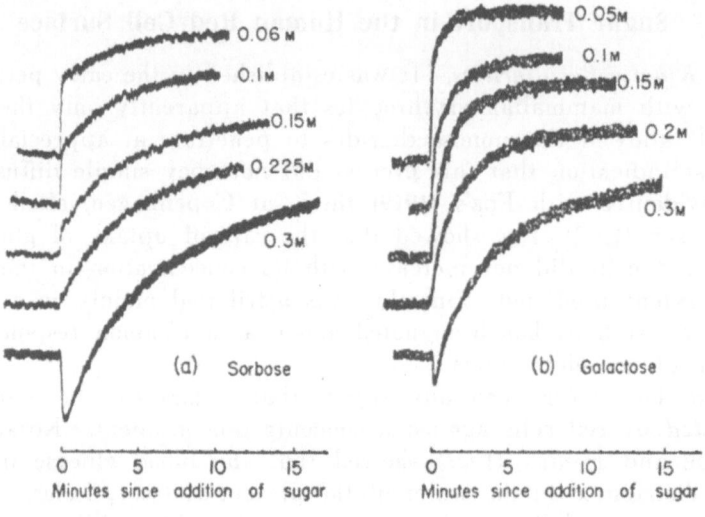

Fig. 21. Kinetics of human erythrocyte sugar uptake as reflected in osmotic volume changes. At zero time, 2 ml. saline medium with sugar at 6 times the indicated final concentration were added to 10 ml. of very dilute cell suspension in balanced saline medium. Downward deflection indicates cell shrinking; upward deflection, swelling; superimposed on this at zero time is upward shift resulting from simple dilution of the suspension by the added solution.
(Courtesy of Journal of General Physiology; from LeFevre and Davies, 1951)

the same transport system, the differences in dynamics being due to differences in avidity for the carrier; it should be noted that the simple rate comparisons do not reveal this relative prowess in attaching to the ferry, which is best brought out in experimental situations involving competition between two or more sugars.

Specificity and Competition.—Wilbrandt (1947) showed that D-xylose and L-arabinose enter the human red cell readily, while their optical antipodes do not; furthermore, these two pentoses interfered with each other's simultaneous entry. LeFevre and Davies (1951) found a competitive interplay between all seven of the monosaccharides previously mentioned (summarized in Table XIV); the least competitive members, levulose and sorbose, proved to be the sugars failing to show saturation kinetics, their sugar-carrier dissociation constants apparently being of a higher order than the sugar concentrations used in the experimental analysis.

Table XIV. *Competition in Movement of Sugars into Human Red Cells.*
(From LeFevre and Davies, 1951.)

In presence of	Inhibition of uptake of						
	Dext.	Mann.	Gal.	Xyl.	Arab.	Sorb.	Lev.
Dextrose . .	—	+++	+++	+++	+++	++++	++++
Mannose . .	+++	—	+++	++	++	++++	++++
Galactose . .	++	-+	—	+	++	+++	++++
Xylose . . .	++	++	++	—	+	++	+++
Arabinose . .	++	++	++	+	—	++	+++
Sorbose . . .	0	0	0	0	0	—	+++
Levulose . .	0	0	0	0	0	0	—

0 — no, or doubtful inhibition
+ — just noticeable inhibition
++ — moderate inhibition
+++ — very marked inhibition
++++ — essentially complete block of uptake

Action of Inhibiting Agents.—Guensberg (1947) noted no change in the glucose uptake upon addition of fluoride, iodoacetate, or insulin, and it was largely on this basis that he sought explanation of the dynamic anomalies elsewhere than in a carrier system. But Wilbrandt (1947) and LeFevre (1947) observed that phlorrhizin impaired the process; moreover, LeFevre (1948) noted that the system was extremely sensitive to Hg^{++} and to the sulfhydryl inhibitor p-chloromercuribenzoate, although not to a broad array of other enzyme inhibitors, including many sulfhydryl inactivators. The effects of the mercurials were prevented or, if caught in time, reversed by addition of sulfhydryl-donating molecules, and LeFevre suggested that the intermediation of a rather unavailable SH-group at the cell surface was an essential step in the passage of glucose into the red cell. Wilbrandt (1950) added gold, chloropicrin, bromacetophenone, and allyl isothiocyanate to the list of effective inhibitors.

Wilbrandt (1950) reported that phlorrhizin or phloretin under certain conditions could inhibit the exit of glucose without delaying its entry, and suggested that these agents act on the step in which the glucose emerges from the membrane, breaking the carrier-complex (presumably involving a phosphatase). By the use of the non-penetrating phloretin phosphate, Wilbrandt and Rosenberg (1950) made this point more specifically. However, LeFevre (1953, 1954) finds the kinetics of inhibition of uptake of the various monoses by phlorrhizin and phloretin to agree with the hypothesis that these inhibitors act by combining reversibly with the carrier, in competition with the sugars. Analysis of the effects of varying the concentrations of inhibitor and sugar have in fact permitted graphical estimation (as in Fig. 22) of the dissociation constants of the carrier-sugar and carrier-inhibitor complexes; the values of these constants (Table XV) are consistent with the differing behaviors of the sugars with regard to the concentration-dependency of their inward transport, discussed above, and were surprisingly similar to those found for intestinal absorption, as cited in an earlier section.

Suggested Models of Carrier System.—In amplifying on the saturation kinetics, Wilbrandt and Rosenberg (1950) noted particularly that increase of the *trans*-concentration (low side of the gradient, whether on outside or inside of the cell) depressed the transfer of glucose much more than would

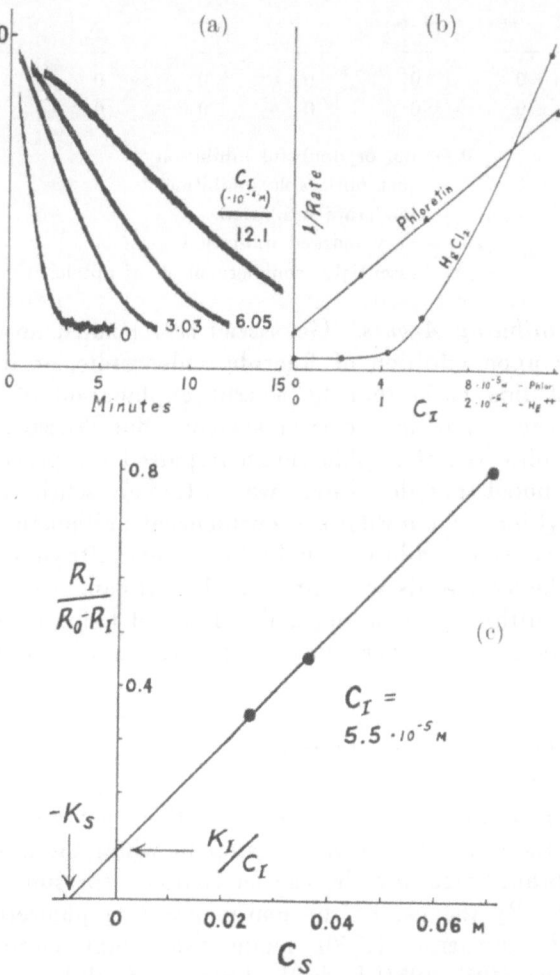

Fig. 22. Inhibition of glucose exit from human erythrocytes, and implications of its kinetic analysis. (a) A 3% cell suspension was equilibrated at 38° C. for one hour with half-isosmotic glucose in 0.7-isotonic saline medium. At zero time, 2 ml. of this was added to 10 ml. of the medium with phloretin at the final concentrations given in the figure. Downward deflection indicates cell shrinkage; records are aligned by backward extrapolation to zero time. (b) The data of (a), and a similar experiment with $HgCl_2$ (non-competitive inhibitor) in place of the phloretin. The reciprocal of the rate of cell shrinking is plotted against the inhibitor concentration. A competitive inhibition is expected to show a rectilinear relation. (c) Procedure as with (a), except that here glucose was added in varying amounts, with or without phloretin at $5.5 \cdot 10^{-5}$ M. R_0 = rate of cell shrinking without phloretin; R_I, with phloretin added; C_s = external glucose concentration. This plot should give rectilinear relation if inhibition is competitive, the intercepts having the significance designated in the figure. K_s and K_I are the equilibrium constants for dissociation of the carrier-sugar complex and the carrier-inhibitor complex, respectively.

Table XV. *Estimation of Red Cell Carrier-Sugar and Carrier-Inhibitor Equilibrium Constants.*

Inhibition of	Sugar K_s	Phloretin K_I	Phlorrhizin K_I
Dextrose entry. . .	0.009 M	0.0000045 M	—
Dextrose exit . . .	0.008	0.0000049	0.000145 M
Galactose entry . .	0.050	—	—
Galactose exit . . .	0.044	0.0000048	—
Sorbose entry . . .	1.3—2.0	—	0.000127
Sorbose exit	2?	0.0000044	—
Levulose entry. . .	2?	0.0000044	—

These figures are tentative, based on only a few experiments reported by LE FEVRE (1954); sorbose and levulose constants cannot be satisfactorily estimated by reason of practical limitations on operating range of sugar concentrations.

be predicted on the basis of simple back-diffusion. They developed a generalized kinetic analysis (WILBRANDT and ROSENBERG 1951) based on the MICHAELIS-MENTEN form of reaction for the transfer in each direction: this would entail a reversible combination with an enzyme at the point of entry into the membrane, followed by diffusion to the other side, and an irreversible decomposition of the complex at the opposite interface. By suitable adjustment of relations between the several constants, this scheme can fit several forms of concentration-dependency, including the types seen with the several monosaccharides. It implies that the individual carrier units operate in only one direction.

LeFEVRE and LeFEVRE (1952) proposed a completely reversible model for glucose transfer which is less specifically defined step-wise; it supposes only, in accordance with the observed saturation behavior in the entry of glucose, that the velocity constants for the reversible carrier-complex formation and dissolution are more rapid at the outer interface of the membrane than at the inner interface, possibly by reason of an enzymatic catalysis. Since it is clear that the glucose-carrier complex is of rather low dissociation constant, the rate-limiting step in this scheme is the dissolution of the complex at the inner surface. This system describes satisfactorily the pattern of movements in all sorts of experimental situations, at glucose concentrations below about ¾ isosmotic, but above this level the movements of glucose are apparently reversibly depressed so that equalization of concentrations may not take place even after many hours. WILBRANDT and ROSENBERG (1950) have suggested that this situation may arise as a result of enzymatic inhibition by a great excess of substrate, as is well known in other systems.

As shown in Table XV, the constants defining the sugar-carrier affinities on the basis of the directly reversible system are of the same general magnitude whether the experimental determination involves entry of the sugar into the cell, or its exit into the medium; also, the constant for a given inhibitor is consistent, with different sugars as the "substrate." Widdas (1953 a, 1954) confirmed many of the observations of LeFevre's group and independently arrived at the same figure for the glucose-carrier equilibrium constant by calculation from the effect of glucose in competing with sorbose. He calculated that the maximal transfer rate corresponded to about a million carrier round-trips/sec. per square micron of cell surface. This is very much less than the number of sugar collisions with the cell surface in the experimental solutions, and only a tiny fraction of the total area need be involved.

Widdas (1953 b) also made the interesting observation that an apparently identical system prevails in the *foetal* red cells of the non-primate mammals (his published studies extend to pig, sheep, deer, and rabbit): hitherto it had been believed that the "permeability" of the primate erythrocytes to hexoses was unique. Assuming a carrier molecule essentially in adsorption equilibrium at both sides of the membrane, Widdas (1952 a, b) has developed a model transport system formally identical with a special case of Wilbrandt and Rosenberg's (1951) scheme; this was originally concocted to handle the behavior of the placental transfer of glucose in the sheep, but is equally descriptive of the red cells' sugar transport system.

Glycerol Transfer in Red Cells

Jacobs (1931), using a hemolytic method, noted wide species variation in mammalian red cells in regard to their penetration by erythritol and glycerol. Moreover, these did not vary in parallel for the two substances, and the high specificities suggested to Jacobs that several special diffusion-facilitating mechanisms must be involved [31]. Schönheyder (1934) found (by hematocrit determinations) that the uptake of erythritol into human red cells followed diffusion law accurately and was not influenced by a variety of experimental factors. Jacobs and Corson (1934) first detected the influence of minute traces of Cu^{++} in retarding glycerol hemolysis in the human and certain other species of red cell; no comparable effect was noticeable with related penetrants such as ethylene glycol. Jacobs, Glassman, and Parpart (1935, 1938) found that the various species could be classified into two categories on this basis, which reflected other characteristics in the process: Cu^{++} sensitivity in the glycerol permeation was associated with a rapid rate of penetration, low temperature-sensitivity, depression by acidity, and (Jacobs and Parpart 1937) depression by alcohols. In addition to man, the sensitive group included rodents but not ungulates and carnivores.

Wilbrandt (1941) found a Cu^{++}-like, but less prominent, effect with Co^{++}, Hg^{++}, Ni^{++}, Zn^{++}, and Ca^{++}. Jacobs and Stewart (1946) noted that, with suitable precautions against interference by extraneous protein, the Cu^{++} effect was evident at concentrations around 10^{-7} M. Even with 90‰ inhibition, the amounts of Cu^{++} required could not have covered more

---

[31] For an excellent general discussion, see Jacobs (1950).

than about 1% of the cell surfaces; furthermore, no diminution of the effect was found when the cells were quickly removed and replaced from a fresh suspension, so that only a small fraction of even this trace of Cu^{++} was actually removed in the inhibitory processes. This suggested a special mechanism for glycerol entry, localized in limited areas on the cell surface.

MELDAHL and ØRSKOV (1940), using ØRSKOV's (1935) photometric method of following osmotic volume changes, found that, unlike the situation with glucose, the "permeability constant" for glycerol and some other penetrants increased gradually during the process of entry, as might be expected from the cell swelling; thus no peculiarity in the kinetics suggested active uptake, as it did with glucose. A number of observations refute the notion that the entry of glycerol in these cells is intimately associated with metabolic activity. Thus HUNTER (1936, 1937) reported that anoxia had no effect on the course of the hemolysis of beef or rat red cells in glycerol and a number of related substances. HUNTER and PAHIGIAN (1940) and HUNTER (1941, 1947 a, b) also found the permeability of chicken and beef cells to glycerol, monoacetin, and several other substances was not depressed by a variety of physical and chemical insults which greatly reduced their respiration and glycolysis, and concluded metabolism had nothing to do with the penetration process.

But LeFEVRE (1947, 1948) found, paralleling the Cu^{++} effect, a somewhat less sensitive response of this system to Hg^{++}, p-chloromercuribenzoate, I_2, and phlorrhizin, and thus suggested that a sulfhydryl group at the cell surface (probably part of a phosphorylating enzyme) took part in the penetration. The sulfhydryl reagents blocked both entry and exit of

Table XVI. *Reversible Inhibition of Osmotic Hemolysis in Glycerol.*
(From LeFEVRE, 1948.)

Inhibitor	Reactivator			Hemolysis time in isosmotic glycerol			
	Identity	Conc.	Time added	Without inhibitor	With inhibitor	With inhibitor and reactivator	
						Total time	Interval
		$M \cdot 10^{-5}$	min.	min.	min.	min.	min.
CuCl$_2$, 10^{-5} M	Glutathione	3	0	1.1	27	1.0	1.0
		4	2			4.4	2.4
	Cysteine	3	0	1.5	19	1.2	1.2
		3	1			2.9	1.9
p-Cl-Hg-benzoate, 10^{-3} M	Cysteine	130	0	2.0	250+	1.7	1.7
			0.3			2.1	1.8
			5			7.0	2.0
			20			24	4
			45			53.5	8.5
			180			200	20

glycerol, and this action was prevented or reversed by addition of cysteine or other SH-donators (Table XVI). Parpart, Barron, and Dey (1947) found chloropicrin similarly effective, but both they and LeFevre tested a great many other sulfhydryl agents which proved to be ineffective; if SH-groups are involved, they seem to be of the most unavailable type, except for the unusual sensitivity to Cu^{++}. No net phosphate movement is involved in the glycerol transport (LeFevre 1948).

As can be seen, differences between the sensitivities of the sugar-transport system and the glycerol-transport system were evident in this work, and it was not suggested that the two overlapped in any way. However, such an overlap appears to be indicated by the remark of Emmens and Blackshaw (1953) that arabinose may prevent entry of glycerol into the red blood cell; unfortunately, the basis for this statement is not given.

Sugar Uptake by Muscles

Older literature gives almost no information about the process of penetration of sugars into muscle cells, in spite of considerable study of the gross phenomenon. The last five years have seen a marked upsurgence of interest in this question, especially in connection with the possible role of insulin in the process. As recently as 1947 (Bouckaert and deDuve), in a review of the insulin literature, the notion that the action of insulin may be primarily by its facilitation of the passage of glucose through the body cell membranes was rejected with little discussion. The change is appreciated in comparing this review with that of Stadie (1954).

Following Gemmill and Hamman's (1941) demonstration in the isolated rat diaphragm of the stimulation by insulin of the glucose metabolism, there has been extensive use of the questionable term "glucose uptake" in further investigation of this phenomenon. Although it now appears that the term is at least partly appropriate here, the first real evidence that the insulin effect involves primarily a translocation, rather than a metabolic conversion, was in the work of Levine et al. (1949). They showed (heavy lines in Fig. 23) that in eviscerated-nephrectomized dogs, the volume of distribution of galactose (a hexose metabolized, if at all, only very slowly) was raised by the administration of insulin (1–2 U/kg./hr.) from 45–47% of the body weight to about 75% (essentially the total body water). It was further shown (Levine et al. 1950) that the addition of insulin after equilibration of the galactose at the 45% distribution level led to a rapid adjustment to the wider distribution level. There was no effect of insulin on the volume of distribution of urea, creatinine, or sucrose, in these dogs and cats. Clearly the insulin acted in some manner to remove a barrier to the passage of galactose at the cell surfaces (mostly muscle in these animals).

However, in measuring the distribution volume or specific activity dilution of C^{14}-glucose, Drury and Wick (1951) found no such effect with massive amounts of insulin in eviscerated-nephrectomized rabbits. Also, elevation of plasma glucose even up to 1% (Wick and Drury 1951 b) gave no increase in the volume of distribution of labelled glucose. Radiosorbitol, which is not metabolized in rabbits, showed a similar distribution volume of about 25%, unresponsive to insulin (WICK and DRURY 1951 a). Moreover. no insulin effect was found with labelled fructose or gluconic acid (DRURY and WICK 1952), but the action on galactose was verified by this method [32]. GOLDSTEIN, HENRY, et al. (1953) duplicated their findings with galactose using L-arabinose and D-xylose, but there was no action of insulin on distribution of D-arabinose or L-rhamnose. There was some degree of

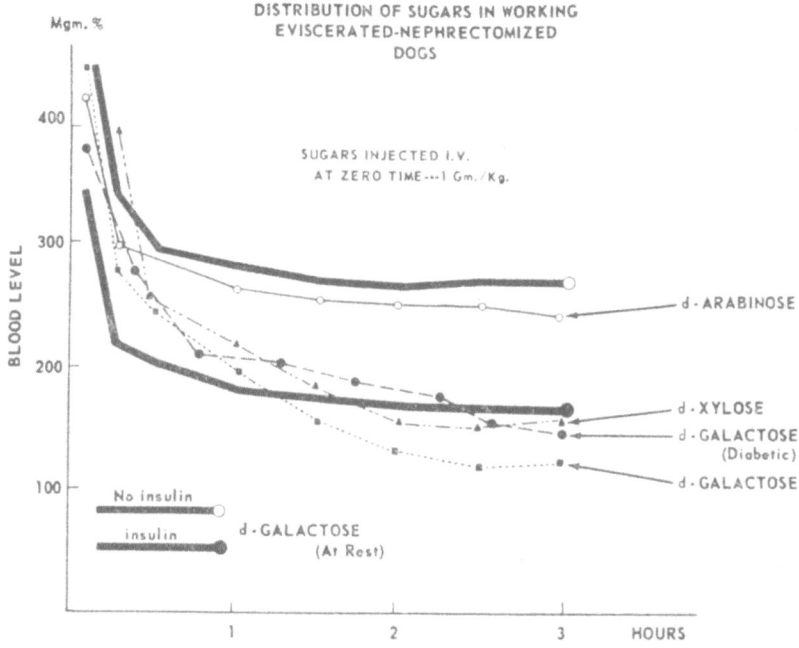

Fig. 23. Effect of insulin and of muscular work on volume of distribution of sugars. Note that the insulin-like activity of muscular work is just as effective in the chronically depancreatized preparation and is thus independent of insulin.
(Courtesy of M. S. GOLDSTEIN and American Journal of Physiology)

utilization of both D-fructose and L-sorbose, so that for these ketoses it was difficult to draw clear conclusions. The summary inference from this work seems to be that the insulin assists intracellular penetration of only those substances in which the first three carbons have the configuration of D-glucose (LEVINE and GOLDSTEIN 1953).

The effect of muscular work in this system was also considered by GOLDSTEIN, MULLICK, et al. (1953). the legs of eviscerated dogs and rats being directly stimulated during the test period. This activity had an effect like that of insulin administration, lowering the equilibrium blood levels of D-galactose, L-arabinose. or D-xylose, but not of D-fructose, L-sorbose. D-sorbitol, D-arabinose, or L-rhamnose (see examples in Fig. 23). It is

[32] Even without insulin. galactose showed a somewhat wider distribution in the rabbit than did the other substances.

postulated that a humoral product of the contracting muscle somehow replaces insulin in activating the transfer mechanism. Wick and Drury (1953 a) extended their earlier C^{14}-galactose studies in rabbits, and found that the insulin-assisted step is depressed by excess carrier galactose or by high blood levels of glucose (probably by a competitive effect). They note that the extrahepatic tissues do maintain a slight galactose metabolism, leading to some $C^{14}O_2$ exhalation in these animals; it is not clear, however, whether this accounts for their observations that, with insulin present, the galactose distribution appears to exceed the entire body water volume. They also showed (Wick and Drury 1953 b) that the increase in glucose utilization in these animals resulting from massive elevation of blood levels was greatly augmented by added insulin. A convenient summary of all this work is given by Drury and Wick (1953).

The validity of the interpretation of the distribution-volume data is perhaps best established by the direct analyses of Park and Johnson (1953) on the tissues of eviscerated rats given a slow continuous intravenous infusion of glucose at 30–1000 mg.%. In the absence of insulin the total organ levels were such that all the glucose may well have been extracellular. With insulin added, the levels in heart and diaphragm were increased 4·5 times, approaching an even-distribution figure; the case was similar, but less decided, in the skeletal muscles.

As noted above, such investigations have thrown new light on the significance of *in vitro* studies (mostly with the rat hemidiaphragm preparation) concerned with the effect of insulin and other hormones on glucose utilization. This subject has been reviewed by Krahl (1951) and by Park (1952). Following Gemmill and Hamman (1941), several groups confirmed and expanded on the effect of insulin in stimulating glucose "uptake" or utilization, and the depression of this process in alloxanized rats: Krahl and Cori (1947); Verzár and Wenner (1948); Krahl and Park (1948); Villee and Hastings (1949); Bartlett, Wick, and Mackay (1949); and Park and Daughaday (1949). Hypophysectomy raised the rate appreciably, and somewhat moderated the response to insulin. Adrenal cortical hormones had little effect on the total glucose utilization, but appeared to depress glycogen formation. Park *et al.* (1952) showed that various preparations of pituitary "growth hormone," given either to the hypophysectomized animal or to its diaphragm *in vitro*, gave a transient boost to the sugar uptake, followed by a prolonged depression, which effect is potentiated by adrenal cortical hormones.

More recently (Bornstein and Park 1953), the interpretation of the rat diaphragm experiments has been greatly assisted by direct chemical analysis of the tissue; it was found that about $^1/_3$ of the muscle volume rapidly comes to equilibrium with extracellular glucose at 200 mg.%. With this system, it was shown that alloxan-diabetic rat serum inhibits the uptake, while insulin prevents this action. The effect of diabetes is lost after adrenalectomy or hypophysectomy; it reappears if the somatotrophic hormone and cortisone are administered *in vivo*, but not if they are given *in vitro* to the muscle. Thus endogenous pituitary and adrenocortical activity are indicated as the basis for the serum principle. PARK (1954) has extended such direct chemical (enzymatic and chromatographic) studies to heart and leg muscles also, and thereby completely confirmed the interpretation of the action of insulin developed above.

STADIE's group at Pennsylvania has made a series of very interesting observations on the combination of insulin and other active factors with muscle, apparently by rigid attachment to the cell surfaces, there regulating the uptake of sugars into the cell. STADIE, HAUGAARD, MARSH, and HILLS (1949) found that rat hemidiaphragms need only a few seconds in the presence of insulin to show the increased glucose uptake and glycogen synthesis at a later time. The effect was already maximal with a one-minute exposure to the insulin at 38° C. (being about 2.5 times slower at 0° C.), and was not responsive to pH over the range 5.7–8.4, or to the availability of oxygen during the insulin treatment. Once attached, the insulin is not readily washed off. If insulin is applied at only 0.1 U/ml., the glucose level in the assay system (0.2–1%) can affect the rate of uptake; but with insulin at 1 U/ml., the glucose level is of no consequence.

STADIE, HAUGAARD, HILLS, and MARSH (1949) looked into the hormonal influences on this insulin attachment. It did not occur in severely hyperglycemic alloxan-diabetic rats; but failed considerably less conspicuously in those cases showing milder hyperglycemia. Adrenalectomy did not alter the diaphragm's ability to combine with insulin, but following hypophysectomy the action of insulin was more striking (STADIE, HAUGAARD, and MARSH 1951 a). Moreover, an extract of beef anterior pituitary gland, given to a rat 20 hours before the operation, depressed insulin fixation. Fractionation of such preparations showed the effective principle to follow the "growth hormone," and purified commercial preparations of this agent showed very strong activity against insulin binding. When the extracts were applied *in vitro* during the attachment process, there was some inhibition in preparations from pituitary or liver and in Cohn Blood Fraction V, but not from muscle or brain preparations. The heat-sensitivity of the pituitary element was like that of the growth hormone, but the purified hormone failed to act *in vitro*. *Pretreatment* of the diaphragm *in vitro* with the active fractions affected the *subsequent* insulin attachment, but this effect was limited to the pituitary preparations.

STADIE (1953) notes that a similar combination of insulin is demonstrable with lactating rat mammary gland cells, adipose tissue, erythrocytes, and leucocytes; only brain cells have so far failed to show the property. STADIE, HAUGAARD, and MARSH (1951 b) showed also delayed metabolic effects of very brief exposures of the diaphragms to epinephrine or to DNP, but not to the adrenal corticoids which have in this system similar actions to these agents.

STADIE, HAUGAARD, and VAUGHAN (1952) directly demonstrated the insulin attachment to rat diaphragm by use of the S^{35}- and I^{131}-labelled hormone. The amount bound was roughly proportional to the concentration supplied up to about 1.5 U/ml.; once attached, the insulin did not wash off even with prolonged exposure to detergents. The total concentration factor could be as high as 6; *in vivo*, still higher concentrations were seen in rat liver and kidney.

Demis (1953) observed that Hg++ very rapidly inhibited the rat dia-phragm's aerobic glucose utilization, while there was some latency in the development of the depression of oxygen consumption (Fig. 24). Moreover, the action on the glucose uptake was readily reversed by BAL, cysteine, or bovine albumin, whereas the effect on respiration was not[33]. Thus it is argued that the action on the glucose consumption is a surface effect and that the inhibited process is a membrane transport system. This inter-

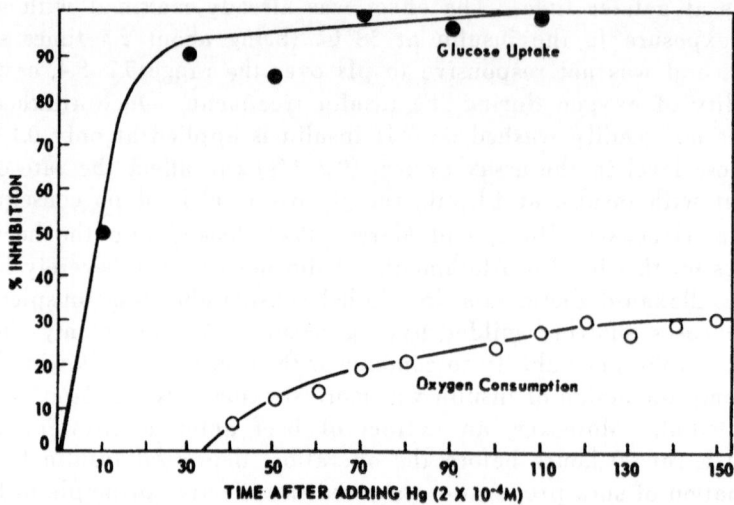

Fig. 24. A comparison of the inhibition by HgCl₂ of aerobic glucose uptake and of respiration of rat diaphragm. Glucose in the medium was measured at 20-minute intervals, and the percentile inhibition plotted represents the average inhibition for the time period extending ten minutes to each side of the point.
(Courtesy of D. J. Demis)

pretation is upheld by the rather rapid respiratory depression by the Hg++ in homogenates of this tissue, indicating that the glucose uptake effect must precede the entrance of appreciable Hg++ into the cells. Moreover, Hg²⁰³-uptake studies and histochemical and polarographic analyses showed an initial rapid binding of Hg++ followed by a slow entrance. Finally, a rabbit antiserum prepared against rat diaphragm hexokinase, which *a priori* would not be expected to get into the cells, slowed the glucose uptake by about 40%. The anaerobic uptake of glucose, galactose, or fructose showed no sensitivity to insulin whatever, although the rates were comparable to those aerobically. At 0° C., the galactose or glucose uptake goes only to about 15% of an equality-distribution. This is still true at 38° C. for galactose at higher concentrations, but at lower concentrations a higher percentage is taken in, so that some sort of saturable active process is implied.

---

[33] Cu++ behaved similarly to Hg++, but the concentrations required for the glucose uptake effect were higher than for the respiration effect.

The same general system seems to prevail also in cardiac muscle. In the perfused rat heart, FISHER and LINDSAY (1954) found that in 30 minutes galactose distributed in a greater fraction of the total volume than did sorbitol or insulin; this apparently intracellular penetration is depressed by increasing the amount of glucose in the medium, and is enhanced by insulin at 0.02–2 U/liter.

Ross (1952) carried out related experiments on a non-muscular tissue with similar results. Entry of glucose into the aqueous humor of rabbits, attached to an automatic glucose injection apparatus maintaining an elevated blood level, was doubled or tripled by one Unit of extra exogenous insulin, and fell to about half with the establishment of alloxan hyperglycemia. As a corollary to this, Ross (1953) kept rabbit eye lenses *in vitro* with glucose at the level found in the aqueous humor; although insulin at $1/3$ U/ml. only slightly augmented the glucose utilization in homogenates of this tissue, it raised the rate in the intact lenses by a factor of 3.5. Moreover, insulin produced an appreciable utilization of galactose, which was not otherwise consumed measurably; and galactose blocked to some extent the uptake of glucose.

Sugar Uptake in Microörganisms

Yeast cells which take up glucose, mannose or fructose at appreciable rates practically exclude galactose, arabinose, and sorbose (ROTHSTEIN and MEIER 1954). On the basis of the characteristics of the initial steps in glucose fermentation by yeast, WERTHEIMER (1934) came to the conclusion that the cell's first reaction with glucose at the surface was a chemical one, not merely a penetration by diffusion. Although WERTHEIMER's evidence on this point was discouragingly indirect, it will be seen that his conclusion has been well established in the last decade. ØRSKOV (1945), using an adaptation to yeast of his red-cell photometric method, noted a suspiciously high rate of uptake of glucose in relation to the yeast's general permeability behavior; however, he found iodoacetate to exhibit no inhibitory action.

ROTHSTEIN and LARRABEE (1948) observed that yeast rapidly removes UO_2^{++} from dilute unbuffered solutions, forming a highly undissociated complex; subsequent uptake is very slow, presumably by a different process. The rapid step saturated at about $6 . 10^7$ U atoms per cell regardless of the concentration in the medium, provided it was high enough to furnish this quantity [34]. But analysis of the cell showed enough phosphates, bicarbonates, *etc.*, to tie up about 100 times as much UO_2^{++} as this. Addition of inorganic phosphate largely reversed the inhibition, at levels much less than the intracellular phosphate level. Thus the marked inhibition of anaerobic glucose consumption which accompanied this attachment was clearly attributable to an action at the cell surface. BARRON, MUNTZ, and

---

[34] This figure is given by BARRON *et al.* (1948) as $7 . 10^6$, but the difference results from an error in BARRON's calculation rather than an experimental difference.

Gasvoda (1948) independently arrived at these same basic findings, and noted that the oxidation of other substrates such as acetate, alcohol, malate. and citrate was not similarly affected by the UO_2^{++}. Rothstein, Meier, and Hurwitz (1951) found the block was also effective against fructose fermentation and adaptive galactose fermentation (but with fewer sites being involved in this case). They observed that although the attachment of the UO_2^{++} was the same anaerobically as aerobically, the glucose utilization was inhibited only about half as much aerobically, extra sites apparently being operative in this condition [35]. Addition of glucose 10 minutes prior to the UO_2^{++} delayed inhibition for some time, the cells evidently meanwhile using up acquired substrate. It was concluded that the UO_2^{++} acts by blocking sites at the cell surfaces normally reacting with hexoses as the first step in their metabolism.

The kinetic analysis of this system was developed by Hurwitz and Rothstein (1951); they showed that the fermentation block was non-competitive, while the respiratory inhibition was a mixture of competitive and non-competitive components. The activation energy of the residual glucose fermentation when inhibited was higher than that of the overall normal process. Rothstein and Meier (1951) concluded from the binding characteristics that cell surface polyphosphates were involved in the UO_2^{++} attachment. Bivalent cations, especially Ba^{++}, Be^{++}, and Zn^{++}, competed weakly with the UO_2^{++}, and it was suggested that their attachment may be a normal and necessary phase of the glucose uptake process. Rothstein and Demis (1953) also observed that the effect of the extracellular K^+ level on glucose (or fructose) fermentation is similar over a range of pH which involves both sides of the K^+-H^+ exchange, so that this too must be a surface effect. Moreover, intracellular $[K^+]$ seems without consequence on the fermentation rates. All this work has been brought together by Rothstein (1954) in a very coherent review.

Glucose Uptake into Brain

Geiger, Magnes, and Dobkin (1953) made the curious observation that in a cat brain perfused with an albumin-sugar-Ringer-red cell suspension. the ability to take up glucose or mannose (whichever was present) was gradually lost, but upon switching to the other of this pair of sugars, the uptake began at full speed. The addition of a liver extract to the perfusion fluid, or inclusion of the liver in the network, cut down the metabolic rate of the brain and prevented the wearing out of the specific up-take processes. Later it was reported that a muscle extract, or the cat's own blood, were also effective (Geiger, Magnes, Taylor, and Veralli 1954). In the blocked brain, elevation of the glucose level was of no avail: lactate accumulated, the electroencephalographic activity failed, and sometimes convulsions developed. Further characterization of the hepatic

---

[35] Rothstein, Frenkel, and Larrabee (1948) showed from the quantitative relation between the binding and the inhibition that about $1/3$ of the attachment sites are not concerned with the inhibition.

principle and analysis of its operation in this system should prove of unusual interest.

Uptake of Polyhydroxy Compounds in Marine Eggs

STEWART and JACOBS (1932 a) found that the process of entry of ethylene glycol into *Arbacia* eggs showed a Q_{10} of about 4 (activation energy of 23,000 cal./mol.) over the range 5 30⁰ C. They also noted (1932 b) that fertilization triples the rate of entry, whereas artificial elevation of the membrane by a brief exposure to distilled water was less effective. The *Asterias* egg was more permeable to the ethylene glycol to begin with, and here the rate was not altered by fertilization.

Transport of Amino Acids

Intestinal Absorption of Amino Acids

The intestinal absorption of amino acids is in general more rapid than that of less significant compounds of similar molecular dimensions, suggesting the likelihood of a special system for facilitating their uptake. CORI (1926 b) observed that the rat absorbed from a mixture of glycine and alanine, or of glycine and glucose, only at a total rate of the same order as if only one ingredient were involved: thus competition was evident even between the different classes of compounds, as if the same mechanism were involved with both. The three ingredients all seemed about equally effective in this competition. However, although phlorrhizin blocked glucose absorption, it failed to alter the uptake of amino acids (WILSON 1932: alanine and glycine, in rats; LUNDSGAARD 1933 b: glycine and glutamic acid, in cats and rabbits; WERTHEIMER 1933: a variety of amino acids, in rats and mice). HÖBER and HÖBER (1937) showed that, in contrast to the situation with polyhydric alcohols or aliphatic acid amides, the percentile absorption of amino acids clearly diminished with concentration, implying a saturable mechanism.

A preferential absorption of the natural forms of a great many different amino acids from racemic mixtures has been observed in dog gut loops (ELSDEN *et al.* 1950; CLARKE *et al.* 1951; GIBSON and WISEMAN 1951); for glutamate and histidine the factor in these experiments was as high as 6. WISEMAN (1951) showed, in an *in vitro* system in which segments of rat gut were perfused intraluminally with a solution initially identical with that on the outside, that the external concentration of the L-form of various amino acids might attain twice the inside concentration, while the D-form remained evenly distributed. Also MATTHEWS and SMYTH (1952) found the venous blood from a tied-off ileal loop in the cat, containing DL-alanine, showed three times as high a level of L-alanine as of D-alanine. WILSON and WISEMAN (1954) found that aerobic conditions were necessary for the uphill transfer of methionine into everted gut sacs.

In a similar (HORVÁTH-WIX) preparation, HETÉNYI and WINTER (1952) showed that glycine and histidine absorption in the rat followed saturation kinetics, faling to increase with the concentration presented at the higher concentrations. Urea, glycinamide, sarcosine, and proline showed no such behavior in the range of concentrations tested; this suggested that both amino- and carboxy-groups are necessary to qualify for the selective absorption. However, since β-alanine behaved like glycine, the spacing of the groups does not seem to be essential. It may be mentioned that the "non-specific" absorptions were as rapid as those deemed "active." However, Schofield and Lewis (1947) found the rates of absorption in rats in the decreasing order: α-alanine, serine, β-alanine, isoserine; and thus concluded that the uptake was slowed by separation of amino- and carboxyl-groups, and by replacing —H by —OH. Klingmuller (1953) reported a very definite preferential intestinal absorption of the D-form of the α-hydroxy acid, mandelate, when a racemic mixture was given.

Wiseman (1953), with an *in vitro* rat gut preparation permitting control of inner and outer solutions, showed definitely that the absorption of the L-isomers could proceed against a concentration gradient. This was demonstrated for alanine, phenylalanine, methionine, histidine, and isoleucine. but not for glutamic acid or aspartic acid.

Renal Reabsorption of Amino Acids

Observations on the specifity in the urinary elimination of amino acids lent an early indication that selective reabsorption processes were involved. A urinary output of nearly pure D-amino acids following injection of racemic mixtures was observed in a variety of mammals by Abderhalden and Tetzner (1935), the natural L-forms evidently being selectively resorbed. This has been better quantified more recently in the cat by Crampton et al. (1951). Doty (1941) pointed out that the tubular reabsorption system handling tyrosine of histidine could not cope with the alkylated derivatives of these amino acids.

Pitts (1943 a, b) showed in intact dogs that the reabsorption of glycine kept pace with the filtration up to a point, then failed at a level corresponding to a tubular maximum of 15–20 mg./min.; saturation of the system with either glycine or alanine or glutamic acid completely inhibited the reabsorption of creatine, but creatine overloading did not prevent glycine reabsorption. Glycine loading also prevented arginine reabsorption. These competition experiments thus suggest that the glycine-carrier complex is a particularly firm one. However, in human renal clearance studies, Ussing (1945) showed that glycine was reabsorbed less effectively than the average amino acid in the natural blood pool. Leucine and valine (taken together) were more effectively absorbed than either glycine or histidine. In Pitts's experiments, there was no interference at the tubules between glucose and glycine (unlike the situation described in the intestine).

Kamin and Handler (1951) have studied the urinary output of endogenous amino acids as affected by a steady infusion of other amino acids. As shown in Table XVII, the output of a given amino acid was generally increased (poorer reabsorption) when the infused acid had similar acidic properties. Particularly effective in depressing reabsorption of other amino acids were glutamine, asparagine and the neutral amino acids; the dicarboxylic acids were particularly ineffective. Threonine, histidine, and especially the acidic amino acids were peculiarly sensitive to the infusion of any extra amino acid.

Table XVII. *Dog Amino Acid Excretion Ratios (with infusion/without infusion).*
(From KAMIN and HANDLER, 1951.)

Infused Acid	Neutral				Basic		Acidic	
	Meth.	Leuc.	Isolc.	Threo.	Hist.	Arg.	Gl. Ac.	Asp. Ac.[1]
Neutral								
Glycine . .	15		28	110	56	7	41	
Alanine . .	22	13	23	81	42	5	620	2500
Leucine . .	64		140	200	55	61	38	38
Methionine .		16	6	270	84	11	83	7
Cysteine . .	13	10	14	110	28	13		160
Tyrosine . .	<1	5	10	25	13	<1	<1	2
Basic								
Histidine. .	6	5	7	76		38	370	5
Arginine . .	3	2	5	19	16		24	6
Lysine . . .	4	2	3	6	5	91	11	3
Acidic								
Glutamic. .	3	3	5	7	5	4		1100
Aspartic . .	<1	1	4	6	2	4	12	
Am. Ac. Amides								
Glutamine .	9	2	2	22	34	4		250
Asparagine .	19	6	12	110	72	5	130	
Average Control Rates (µg/min.). . .	1.4	5.2	1.8	2.6	4.1	3.8	5.0	<0.5

[1] Figures in the last column only are in terms of µg/min., since the control excretion rate is so low that the ratio cannot be accurately expressed.

BEYER *et al.* (1946) demonstrated that the capacity of dog kidneys to reabsorb tryptophane, isoleucine, valine, and leucine was far in excess of the normal load, so that artificial increase of the plasma concentrations up to seventy times normal did not lead to appreciable loss in the urine, percentagewise. The same group later showed the same behavior with threonine and phenylalanine (RUSSO *et al.* 1947) and with histidine and methionine (WRIGHT *et al.* 1947); on the other hand, arginine and lysine showed thresholds at a filtration rate of somewhat over 10 mg./min. When the acids were given in pairs, competition was apparent between any two acids within the groups: arginine-histidine-lysine and leucine-isoleucine, but not between the one group and the other (Beyer *et al.* 1947). Thus two independent reabsorptive pathways were indicated, and glycine was evidently handled by yet another system.

Mudge and Taggart (1950 a) reported that while DNP, at 0.2 mM in the plasma, greatly depressed the dog's tubular transport of PAH and PSP, it did not alter reabsorption of glucose or glycine. DNP, and even more so TNP (picric acid) and 2, 4-dinitro, 6-phenyl phenol, also stopped accumulation of phenol red in the tubules of teased-off fragments of flounder kidney (Taggart and Forster 1950), apparently indicating a role for aerobic phosphorylation in the process. A number of related substituted phenols were not active in this system, and these also failed to stimulate the respiration in a kidney mince. Accumulation was also depressed by anoxia or chilling, and by a long list of metabolic inhibitors (Forster and Taggart 1950; Beyer *et al.* 1950). Mudge and Taggart (1950 b) found that acetate or lactate enhanced the PAH transport, while succinate or fumarate were inhibitory. This exactly paralleled the *in vitro* behavior of rabbit kidney slices (Cross and Taggart 1950). Shideman and Rene (1951) observed that a group of inhibitors preventing excretion of PAH and PSP in the intact dog, or their accumulation in rabbit or guinea-pig kidney slices, also showed *in vitro* inhibition of succinoxidase; such agents in general did not affect absorption of glucose or phosphate.

Amino Acid Uptake from Blood to Tissues

Inequality of Normal Distribution.—Van Slyke and Meyer (1913) showed by direct chemical analysis for total free amino acids that, thirty minutes after dogs received intravenously a casein hydrolysate, their tissues had 5—10 times the levels found in the blood; the liver and kidney showed particularly high increments over the previous levels. But Ussing (1943 b) was unable to show any such concentrative ability in guinea-pig liver, kidney, muscle, or brain, with respect to leucine, valine, or tyrosine. Ussing believed technical criticisms cast serious doubt on claims of this sort, with the exception of Cohen's (1939) study on glutamic acid distribution in various tissues of sheep, guinea-pig, and pigeon, in which high intracellular excesses were shown to be quite general.

Recently a group at Harvard and Tufts has been especially active in study of the uptake of amino acids by a variety of types of cells. Feeding guinea-pigs a meal of a single amino acid, and determining the muscle, liver, and plasma levels three hours later, Christensen, Streicher, and Ellinger (1948) found the same general distribution noted by the earlier workers (liver/muscle/plasma ratio for glycine was about 33/9/1). Feeding any other of a long series of amino acids raised the levels in liver and muscle less than the plasma level, so that the concentration factor was lowered; and concomitantly, the glycine distribution was altered in the same manner. Only glutamic acid feeding increased the cell/plasma concentration ratios, not only of itself, but also of glycine and of the remaining acids (analysed together); this meant an actual decrease in the plasma levels.

In late fetal guinea-pigs, the plasma and tissue levels were about three times the maternal levels (CHRISTENSEN and STREICHER 1948). Feeding histidine, methionine, or proline to the mothers greatly reduced the fetal/maternal ratio of plasma glycine; but glutamic acid fed to the mother was heavily concentrated into the fetal plasma (without involving any alteration of the glycine distribution in this instance).

Association of Uptake with Metabolic Factors.—The rat hemidiaphragm preparation also was used by this group for in *vitro* studies of amino acid distribution (CHRISTENSEN and STREICHER 1949; CHRISTENSEN, CUSHING, and

Table XVIII. *Maintenance of Glycine in Rat Diaphragm Cells* in vitro.
(Adapted from CHRISTENSEN and STREICHER, 1949.)

| Medium | Glycine, as mg. % of N | | | | % Change in Distrn. Ratio |
| | Plasma (x 1.05) | Fresh Diaphragm | After Incubation | | |
			Medium	Diaphragm	
Rat plasma	0.42	1.90	0.52	1.78	— 29
Krebs-Ringer-bicarbonate .	0.55	2.4	0.53	1.8	— 22
+ KCN, 3 mM	0.47	2.03	0.66	1.4	— 51
+ Na pyruvate, 20 mM.	0.34	1.50	0.37	1.80	4
+ Na citrate, 15 mM .	0.46	2.13	0.46	1.56	— 26
Pyruvate medium (3 hrs.).	0.63	2.04	0.43	1.59	14
+ DNP, 10^{-4} M	0.24	1.28	0.52	1.11	— 60
Rat plasma + pyruvate, 20 mM	0.36	1.33	0.32	1.39	16
in place of pyruvate:					
Succinate					— 13
α-ketoglutarate					3
Fumarate					— 17
Glucose					— 29
Rat plasma + pyruvate, + Arsenite, 25 mM . .					— 56
+ DNP, 10^{-5} M. . . .					— 16

One-hour incubations except where otherwise noted; opposite hemidiaphragms used as controls.

STREICHER 1949). As Table XVIII shows, the normal distribution of glycine (intracellular/extracellular ratio of about 5) was maintained in *vitro* at 37° C. for some hours with pyruvate, α-ketoglutarate, or succinate at 20 mM in the medium, and glycine at about the same level. Varying the level in the medium led to lesser variations in the same direction within the tissue. The temperature coefficient of the leakage outward was on the order of that for a diffusion process; since elevation of temperature favored higher tissue concentration, the entry process apparently has a definitely higher temperature-dependency. The cellular uptake was depressed somewhat by anoxia, cyanide, arsenite, DNP, or excessive [K+], some of which are represented in Table XVIII.

A considerable glycine concentration can also be maintained by duck or human red blood cells [36] at 38° C. *in vitro* (Christensen, Riggs, and Ray 1952) and by rabbit reticulocytes, but not by mature rabbit erythrocytes

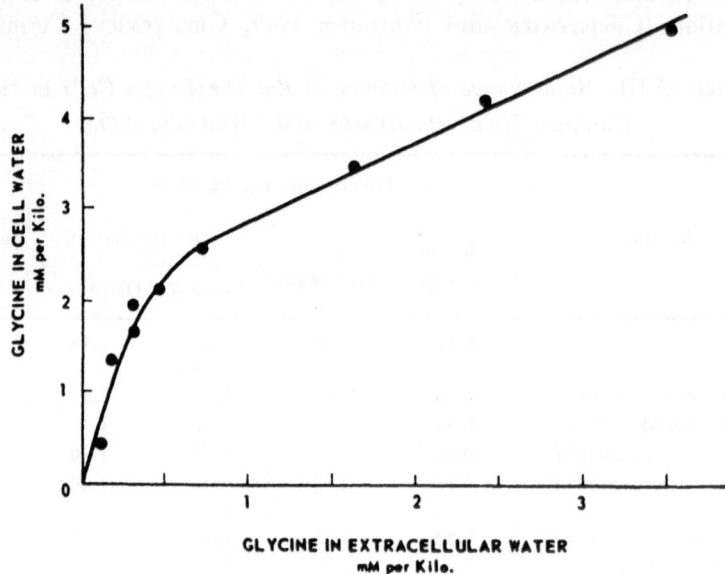

Fig. 25. Relation of glycine level of duck red blood cells to level in the medium. 2 hrs. incubation at 38°C. in atmosphere of 5 CO$_2$: 95 O$_2$.
(Courtesy of H. N. Christensen and Journal of Biological Chemistry)

(Riggs, Christensen, and Palatine 1952). In the duck cells, it was shown that although the concentration ratio fell with increasing concentrations of glycine, the *gradient* was maintained over a wide range, at about 1.4 mM (Fig. 25). The process in the duck and human proved insensitive to anoxia, cyanide, DNP, arsenate, fluoride, Cu++, or phloretin phosphate, but was depressed by high [K+]; curiously, in the light of the above, the system in the rabbit reticulocytes was depressed by fairly high concentrations of DNP, arsenate, or cyanide. Human cells concentrated also L-glutamic acid and L-alanine, but not D-alanine, whereas duck cells accepted both alanine stereoisomers equally readily. Klingmuller (1953) noted that red

---

[36] Johnson and Bergeim (1951) contrast an array of amino acids with regard to their susceptibility to accumulation by human red cells. Earlier, Ussing (1943 a) had questioned such observations, attributing the high red cell amino-N largely to glutathione.

cells (species unnamed) do not distinguish, as does the intestine, between
D- and L-mandelate.

In the squid giant axon, KOREY (1950) found (with N^{15}-labelling) that
the passage of L-aspartic acid was only about ¼ as rapid as the passage
of glycine; but it might be concluded from studies on rabbit brain slices
(KOREY and MITCHELL 1951) that aspartate enters these cells by an active
process (glycine entering passively). The Q_{10} for aspartic acid uptake was
higher than for glycine [37], and only the aspartate entry was depressed by
cyanide. Acetylcholinesterase inhibitors proved ineffective on this system.
Accumulation of L-glutamate by guinea-pig cortex slices, on incubation
with glucose, was described by STERN, EGGLESTON, HEMS, and KREBS (1949):
the tissue level of glutamate can build up to ten times the concentration
in the medium in this system, even though the consumption rate is considerable. Fructose, pyruvate, lactate, and F-1,6-diP can substitute for
glucose as the substrate, but a number of other carbohydrate intermediaries
cannot.

Test of a wide variety of inhibitors showed that the uptake process and
the utilization process had entirely different sensitivities. Fluoride,
phlorrhizin, crystal violet, DNP, and iodoacetate depressed or even abolished
the uptake process. Kidney cortex, spleen, lung, and chorion also accumulate glutamic acid, but this does not apparently require glucose; a wide
variety of other tissues did not appear to concentrate glutamic acid at all.
SCHWERIN, BESSMAN, and WAELSCH (1950) have particularly contrasted
various rat and mouse tissues in regard to their glutamic acid and glutamine uptake.

The free cells of the mouse EHRLICH ascites carcinoma have provided
another convenient system for study of these processes. The ascitic fluid
in the tumor-bearing mice contains amino acids only at or below plasma
levels, while the cells maintain concentrations which exceed 10 times these
levels in the case of glycine, glutamine, and glutamic acids (CHRISTENSEN
and RIGGS 1952). This concentrative activity was further raised during
in vitro incubation at 38^0 C.[38] The activity was little depressed at 20^0 C..
although it was essentially abolished at 1^0 C. or 51^0 C. Anoxia, cyanide,
DNP, iodoacetate, fluoride, arsenate, or antibiotics depressed the concentrative action, but many other inhibitors did not, notably azide and
phloretin phosphate. Altogether (CHRISTENSEN, RIGGS, FISCHER, and PALA-
TINE 1952 a), 17 amino acids where shown to be accumulated, some of these
being rare or even utterly unnatural substances; in general, L-forms were
selected over the D-forms.

Competition and Specificity.—In the rat diaphragm work, it was found
that the glycine concentration ratio fell 20—40% in the presence of equal
concentrations of any of a long series of other a-amino acids (or mixtures

---

[37] Although the data do not actually support the figures given in the report,
they do appear to justify this qualitative statement.

[38] A gradient of 50 mM was maintained over a wide range of glycine concentrations in the medium (CHRISTENSEN and HENDERSON 1952).

thereof), but not with any of the β-amino acids: N-formylation of valine abolished its effectiveness, but similar N-alkylation of other α-amino acids had no such action. D- and L-forms of a given acid were equally effective (this might be expected, since glycine is optically inactive).

Similarly, in the ascites carcinoma cells, glycine accumulation was depressed by addition of alanine (but not of glutamate), and was actually stimulated somewhat by lysine, as shown in Fig. 26. Dipolar acids inhibited the uptake of other dipolar acids, the competitive prowess

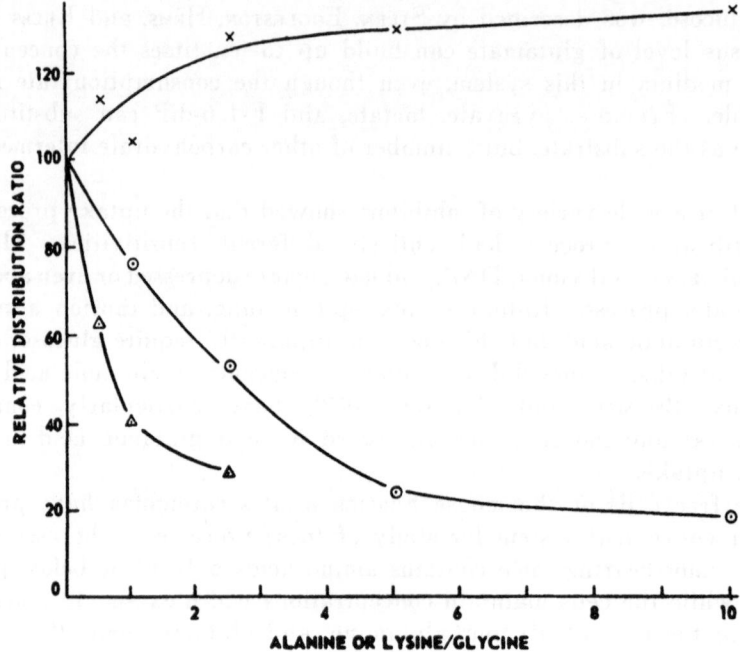

Fig. 26. Glycine concentration in Ehrlich ascites tumor cells as affected by alanine or lysine. Circles — glycine level fixed; varied alanine level. Triangles — glycine level varied; alanine level fixed. Crosses — glycine level fixed; lysine level varied.

(Courtesy of H. N. Christensen and Journal of Biological Chemistry)

correlating with the individual capacities to be concentrated. Uptake of dipolar acids was not affected by the anionic acids and was generally enhanced by the cationic acids, whose own uptake was depressed by other cationic acids and by the dipolar acids. In comparing the acids' uptakes, it was brought out that higher concentration was favored by shorter chain lengths, and by presence of polar groups on the side chain. The structural factors in the process were given special attention in a later report (Riggs, Coyne, and Christensen 1954), largely by testing of various relatives of tryptophane. Removal of the amino group greatly reduced the uptake, while tryptamine (missing the carboxyl group) was concentrated to a moderate degree. Methylation of the indole-N moderately depressed the concentrative action. Similarly, glycinamide, N-acetyl glycine, and a number of diamines, aminobenzoates, and pyridine carboxylates were rather poorly taken up. In an aromatic molecule, a 1,2 or 1,3 relation between amino- and carboxy-groups is unfavorable to accumulation. A *second* nitrogenous group is especially favorable, even on a ring structure; a separation of 2 N atoms by 3 carbons is better than a separation by 4 carbons. It was apparent that the *spacing*, rather than the *position*, of a chain on a pyridine ring, was the critical factor in this. It is thus suggested that the NH_2-group is the site attacked in the uptake process. The concentrative uptake is not limited to the α-forms (CHRISTENSEN, HESS, and RIGGS 1954): taurine, β-alanine, and triiodothyronine are also concentrated appreciably.

Role of Cations and Other Accessory Factors.—A certain amount of K^+ in the medium was stimulatory, but at 40 mM or higher, K^+ was distinctly depressing. K^+ movement proved to be involved in the uptake of the heavily concentrated diamino acids, especially a, γ-diaminobutyric. Entering as cations from a 30 mM level in the medium, these acids displaced nearly all the cell K^+ within an hour and then proceeded to accumulate, partly in exchange for cell Na^+ and partly by bringing in Cl^-, so that the cells swelled considerably by osmosis (CHRISTENSEN, RIGGS, FISCHER, and PALATINE 1952 b). Mg^{++} however was never extruded by this process. No similar behavior was observed with canine or human red cells or with rat diaphragms. However, when one of these acids was injected into the carcinomatous mouse, the cancer cells had acquired little of it at a time when half of the liver K^+ had been already replaced thereby. Muscle and kidney also received appreciable shares, in exchange for K^+.

RIGGS, COYNE, and CHRISTENSEN (1953) studied the enhancing action on this system of pyridoxal; this effect is pronounced over a limited band of concentrations in the neighborhood of 1 mM. No similar effect was seen with any of several relatives of pyridoxal or a variety of B-vitamins. Extensive *pre*treatment with pyridoxal was ineffective; the agent's presence in the medium at the time of the concentrative activity was necessary. Since pyridoxal phosphate as a coenzyme is generally supposed to form a chemical linkage with amino acids, it was suggested that it thus "grasps" the acids as a part of the transport process (although the phosphate is if anything less effective here than the plain pyridoxal). If the agent does act as a carrier, it is not consumed in the process, since the incremental induced uptake amounted in some instances to five times the number of molecules of pyridoxal added to the medium.

The application of pyridoxal also induces a loss of K^+ and a smaller gain of Na^+ (with cell shrinkage), in both the Ehrlich tumor cells and human erythrocytes (CHRISTENSEN, RIGGS, and COYNE 1954), and certain complexities in the relations between the ionic and amino acid movements are brought out in this report. This recalls the evidence relating glutamate uptake and cation maintenance in nerve cells.

Amino Acid Uptake by Microörganisms

Gale (1947 a) described similar complications in the pronounced accumulation of free amino acids in *Streptococcus faecalis*. Lysine can accumulate in these bacteria without any apparent special exogenously supported activity (Fig. 27); but glutamine, glutamic acid, and histidine require metabolic assistance, as from glucose fermentation (which depresses the uptake of lysine). The pH-dependence of the rate of entry suggests that lysine moves through the membrane largely in the isoelectric form, as classical permeability concepts would predict, but the optimal pH for glutamic acid entry depends on the substrate being used. Furthermore,

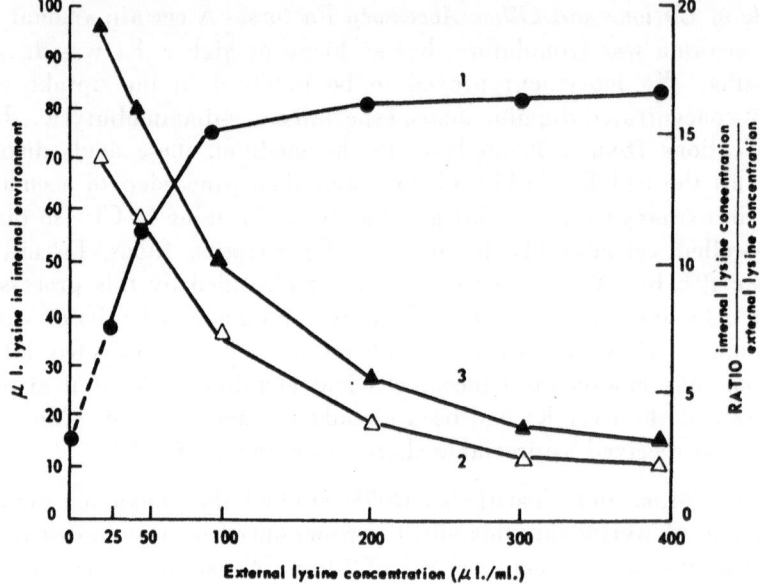

Fig. 27. Uptake of lysine by *Streptococcus faecalis* as a function of its level in medium. Cells suspended for 3 hours at 4°C. in lysine solutions. 1. Internal lysine concentration. 2. Concentration ratio based on intact cell volume. 3. Concentration ratio based on calculated cell water volume. Amino acid is expressed as μL CO_2 released by specific decarboxylase in the manometric analytic procedure.

(Courtesy of Journal of General Microbiology; from Gale, 1947 a)

the Q_{10} for lysine entry is only 1.4, while that for glutamic acid is about 2.3. Once captured, glutamic acid and lysine both require fermentation to escape again, however. Some degree of mutual concern with a common pathway is also indicated by the depression in the presence of either amino acid of the uptake of the other.

Taylor (1947) extended these observations over 27 types of bacteria and yeasts and found that the 16 Gram-positive forms (including three yeasts) also accumulated glutamic acid and lysine against a gradient, while the 11 Gram-negative forms consistently failed to do so. Taylor (1949) noted that glucose fermentation can provide the energy for glutamic acid accumulation also in yeast cells.

Gale and Mitchell (1947) found that prior or simultaneous exposure to triphenylmethane dyes (which interfere with glutamate metabolism) enhanced the uptake of glutamic acid; this effect was more pronounced in the alkyl-substituted compounds, correlating with the dyes' relative lipid solubilities. The action of penicillin on the system has proved especially interesting; GALE and TAYLOR (1947) observed that the intracellular glutamic acid level fell, following a lag period, in *Staphylococcus aureus* exposed to this antibiotic, although there was no change in rates of respiration, glucose oxidation, glucose fermentation, or lysine accumulation. There was also no disturbance of the internal glutamic acid metabolism, so that the acid continued to disappear after the penicillin inhibition of its entry

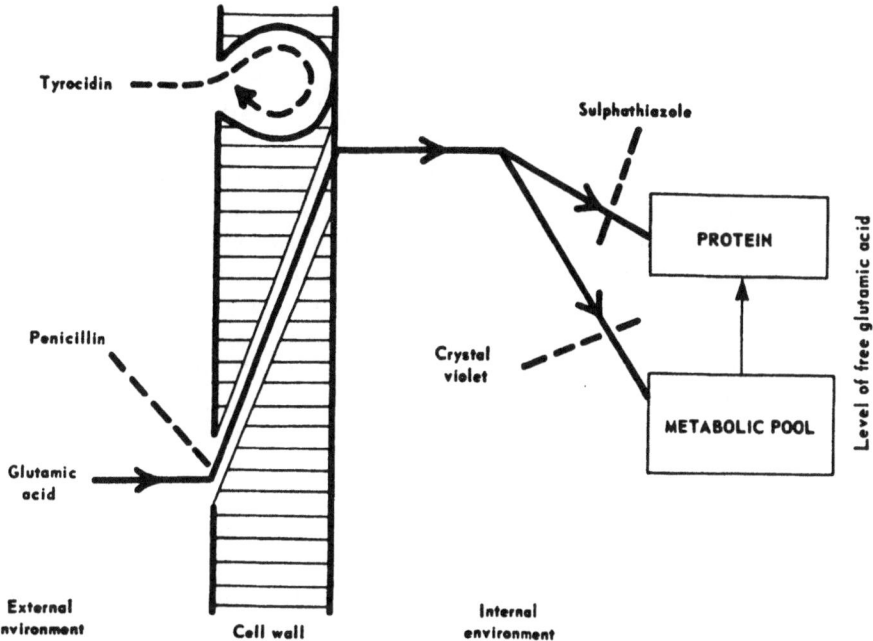

Fig. 28. Interpretative scheme of action of various agents on assimilation of glutamic acid by Gram-positive bacteria.

(Courtesy of Journal of General Microbiology; from GALE, 1947 b)

took effect. It was later noted (GALE and RODWELL 1948, 1949) that *Staph. aureus* strains which had developed penicillin resistance had lost the power to concentrate glutamic acid, and survived on NH_3, thiamine, and glucose, without amino acids; and that by training ordinary *Staph. aureus* toward this condition by gradually diminishing the amino acid supply, increased penicillin resistance can be developed. PAINE (1951) found that bacitracin, if given some time in advance, blocked glutamic acid accumulation in *Staph. aureus* without any effect on fermentation or respiration. Sulfathiazole enhanced the accumulation, apparently by blocking condensation of the glutamic acid into peptides (GALE 1947 b). The summary scheme of the action of the various inhibitors is given in Fig. 28.

Certain divalent cations are also apparently important to the function of this glutamate uptake system (GALE 1949); Mn++ is superior for reactivating washed cells, while Mn++ is more effective in maintaining the function during rapid growth. Mn++ and other divalent cations are also capable of counteracting inhibition of the system by the chelating agent 8-hydroxyquinoline.

The uptake of glutamic acid in the staphylococcus is depressed by DNP at concentrations far below those inhibiting glucose fermentation, according to GALE (1951 a). Azide acts similarly, but less strikingly; while fluoride inhibition of fermentation parallels that of the accumulative process. The pattern in *Strep. faecalis* was entirely different: DNP or azide greatly enhanced glutamic acid accumulation, at the same concentrations which are inhibitory in *Staph. aureus.* Crystal violet and arsenate also augment the uptake in the streptococcus, and it is suggested that the effect with all the agents is attributable to a block of metabolism of the amino acid (so that it piles up more rapidly) rather than to a true stimulation of the uptake process.

In contrast to the cation displacement reported in tumor cells by CHRISTENSEN's group, DAVIES, FOLKES, GALE, and BIGGER (1953) observed that uptake of glutamic acid in *Staph. aureus, Strep. faecalis,* or *Saccharomyces fragilis* is accompanied by an uptake of K+ as noted in an earlier section (see Fig. 8); this amounts to slightly less than 1 atom of K per molecule of glutamic acid.

GALE (1953, 1954) has succinctly reviewed all this material and correlated it with related findings of other groups. He notes that comparison of C^{14}-labelled amino acid accumulation, with and without the inhibitor DNP, reveals the following sequence in regard to the significance of the active process:

proline ⟩ glutamic acid ⟩⟩ phenylalanine ⟩ aspartic acid, methionine,
     valine ⟩ lysine ⟩ threonine ⟩ arginine, alanine, glycine, tyrosine

the last group showing no sensitivity to DNP at all, but still building up to high internal/external concentration ratios. Glutamic acid can also be taken up from various esters, and even simple peptides thereof.

GALE (1951 b) concludes that, although some of the acquired glutamic acid in *Staph. aureus* is metabolized, none is combined to form higher molecules, unless a mixture of other amino acids are also available. Under the latter conditions, the free glutamic acid in the cells falls rapidly as it becomes fixed in chemical combination. However, in the 1953 review, GALE was inclined to the view that even the "free" acid may well be held in a highly labile, but non-diffusible form.

Transport of Fatty Acids
Intestinal Absorption of Fats

The problem of the mechanism of fat absorption in the intestine is more complex than that of the mechanism of absorption of the water-soluble foodstuffs discussed in earlier sections. The bulk of the work on fat absorption, in fact, has been concerned with the physicochemical state of the absorbed material and the role of the digestive preparation for absorption, or with the anatomical fate of the material after its primary passage through the barrier. Neither of these matters is directly relevant to the present subject. The analysis of the role of chemical processes in the epithelial cells in the trans-mucosal movement has been given less attention. VERZÁR and his associates have presented considerable evidence, however, of the importance of phosphorylation in the absorption.

Oleic acid was more rapidly absorbed in the rat if glycerophosphate, or both glycerol and phosphate, were present in the lumen (VERZÁR and LASZT 1934 a). Subcutaneous injection of iodoacetate stopped the fat absorption (VERZÁR and LASZT 1934 b), but this treatment is so rough on the animals that the significance of the observation is questionable. The reversible depression of absorption of olive oil or animal fat by phlorrhizin (VERZÁR and LASZT 1935 a) is more indicative, but the objection may be made that the doses required were tremendous.

The absorption of fat in the rat is decidedly inhibited a few days after adrenalectomy (VERZÁR and LASZT 1935 b). However, BARNES et al. (1939) found that salt-maintenance prevented this, and they attributed the effect to a secondary disturbance resulting from the poor condition of the animals following adrenalectomy. But BAVETTA et al. (1941, 1943) could not duplicate BARNES's observations; they noted that the effect of adrenalectomy was restricted to the longer-chained fatty acids.

There had been considerable older work suggesting that phospholipids were formed in the course of fat absorption, but these observations were in terms of altered blood and tissue contents of lecithin and related compounds and did not show that this had anything to do with the transfer processes. SCHMIDT-NIELSEN (1946) verified in rats REISER's (1942) observation in swine that there was no increase in total phospholipid during fat absorption. However, with P^{32} present in the animal's circulation, it was shown that the turnover into gut phospholipid was considerably augmented during the absorption of oleic acid. Phlorrhizin did not prevent this. No such change accompanied a similar degree of general absorptive activity with respect to glucose. Thus the phospholipid formation appeared to be a part of the cell transport process in the fat absorption; it may be noted, however, that the specific activity of the phospholipid never approached that of the total gut wall inorganic phosphorus, so that joint phosphate-lipid membrane transfer is not implied.

COLLET and FAVARGER (1951) showed, with H^2-labelled glycerol, palmitic, and elaidic acids, a much more rapid turnover into the mucosal phospholipids of fatty acids than of glycerol, during absorption of the mixture, in monkeys and rats.

Although fatty acid permeability has been studied in a number of cell types, there does not seem to be special evidence of metabolic dependence or like complication in the permeation processes, other than in intestinal absorption.

Transport of Water

Intracellular Tonicity; Extrusion of Water from Mammalian Cells

It is generally appreciated that vertebrate tissues do not readily maintain their natural volumes when removed from the body to media resembling fairly closely the body fluids and having the same osmotic pressure as the blood plasma; some degree of swelling is the general rule. For instance, Sperry and Brand (1939) describe the swelling of rat liver slices at room temperature in saline, Tyrode's, or Krebs's solutions. Amounts of water equalling the dry weight of the tissue were taken in in about ten minutes, and again in the next fifty minutes.

The earlier literature on this point is well reviewed by Robinson (1953); much of this is complicated by the use of pure salt or sugar solutions instead of balanced media. Elliott's (1946) observations on the behavior of rat cerebral cortical slices show the typical patterns: the slices swell rather rapidly to a new steady state (which is maintained for some time if the tissue is surviving) in isotonic or even hypertonic NaCl, and this is still worse in KCl or non-electrolytes; small amounts of salts greatly reduce the swelling in sugar solutions; addition of serum or high concentrations of gelatin cuts down on the swelling but does not abolish it.

Opie (1949) found that slices of rat liver parenchyma or kidney required NaCl solutions of about twice the osmotic pressure of the blood plasma to prevent swelling; for pancreas slices, the figure was still higher, while for some other tissues it was not far above "isotonicity." Evidence that this behavior was probably not the result of post-dissection changes in the cells was added (Opie 1950) by the demonstration that similar swelling was not seen in the liver and kidney cells following poisoning of the animals with $HCCl_3$, CCl_4, or K_2CrO_4, and that recovery from such poisoning led to restoration of the vital osmotic imbalance. Opie suggested that the basis for the phenomenon is that the cell contents *in situ* are normally maintained at a higher osmotic pressure than the extracellular fluids.

Supply of adequate oxygen assists in limiting the swelling (see for instance Stern *et al.* 1949), but Robinson (1950 a, 1952 a) found that even in solutions which were sufficiently well-balanced that they supported a constant respiration for some hours, rat renal cortical or liver slices would swell unless the normal metabolic reaction patterns were functioning. At $0–4^0$ C., the water content of the slices exceeded *in vivo* figures even at tonicities of 0.58 osm (about twice that of plasma); cyanide or DNP (Robinson 1950 b) produced the same effect at body temperatures. But if the slices were kept respiring normally at 38^0 C., not only was the water content kept down toward *in vivo* levels, but even the anticipated response to *hypotonicity* was greatly reduced. The adjustments in any particular solution were evidently made in the first two minutes of exposure. All this clearly suggested that the osmotic pressure of the cells was much higher than in the medium, and that respiratory activity (phosphate-bond dependent), rather than the osmotic pressure, was the chief factor in determining the water content. The energy of respiration proves to be ample to take care of the necessary active water-extrusion this notion would entail.

The direct demonstration or disproof of such intracellular hypertonicity is not an easy technical problem. The early observations of SABBATANI (1901) on the freezing point of various animal tissues showed certain organs (notably kidney and liver) to have an unusually high, and unusually variable, apparent osmotic pressure. Chemical analyses (ROBINSON 1952 b) showed in rat renal cortex slices a considerable elevation of the doubly misnamed "fixed base" over the level in the plasma or the medium in which the slices were kept actively respiring. This difference was abolished by cyanide or by chilling.

However, CONWAY and McCORMACK (1953) came to different conclusions; with a special microcryoscopic technique, they carried out a thorough study on the freezing-point depression in freshly ground bits of rat and guinea-pig tissues. When ground at room temperature, many tissues (especially rat diaphragm) showed much greater freezing-point depression than found for plasma. But if frozen first, ground, and tested rapidly, little deviation from the plasma figure was found; the deviation increased progressively within several minutes at 0° C. Extrapolating back to the instant at which thawing occurred, the freezing points essentially coincided with that of the plasma. Azide or iodoacetate had little effect on the apparent rapid rise in osmotic pressure, but $HgCl_2$ inhibited the process for a relatively long time, so that even the rat diaphragm preparation did not appear hypertonic to plasma.

AEBI (1950) analysed the *in vitro* behavior of guinea-pig liver slices in terms of maintenance of N-content, volume, and overall respiration, as a function of the ionic make-up of the medium. He found Ca^{++} the most significant component in avoiding swelling and loss of N, and K^+ most significant in maintaining the respiratory rate. Optimal conditions were with normal serum levels of Na^+, Ca^{++}, and Mg^{++}, but about double the serum level of K^+. Still higher levels of K^+ (up to about 50 mM) were required to keep the swelling down to a few percent. of the original volume (AEBI 1951). Addition of various colloids at fairly high concentrations to Na^+-media also minimizes swelling, but does not avoid loss of K^+ (AEBI and MEYER 1951).

AEBI (1953) pointed out that even under the most favorable circumstances *in vitro*, in which the water content of the tissue slices is maintained and K^+ is accumulated (or at least not lost), the Na^+-extrusion system partly fails, so that the Na^+ content of the tissues continues to rise. He expresses the need for a more direct demonstration of intracellular hypertonicity than is given by this type of experiment, which admits of other interpretations.

DEYRUP (1953) made the interesting observation that the swelling of rat kidney slices, although more pronounced in monosaccharide solutions than in salt solutions, does not occur in isosmotic solutions of the disaccharides sucrose, maltose, or lactose (Table XIX). In fact, if such a slice was transferred from Krebs-Ringer-phosphate to a sucrose solution of like osmotic pressure, a pronounced re-shrinkage resulted. Deyrup suggested therefore that Robinson's interpretation of an intracellular hypertonicity

Table XIX. *Changes in Relative Water Content of Rat Renal Cortical Slices Immersed in Various Media for Thirty Minutes.*
(From Deyrup, 1953.)

Medium	Percentile Change in Relative Water Content (Av. ± standard deviation)	No. of Expts.
Heparinized rat blood	24.4 ± 15.8	4
Krebs-Ringer-phosphate, pH 7.1 . . .	29.4 ± 18.5	22
NaCl, 150 mM	45.9 ± 27.6	18
NaNO₃, 150 mM	61.5 ± 31.6	6
Glucose, 0.3 M	42.4 ± 13.5	11
Fructose, 0.3 M	45.8 ± 28.6	6
Galactose, 0.3 M	57.9 ± 31.4	6
Sucrose, 0.3 M	− 2.4 ± 17.3	33
Maltose, 0.3 M	− 19.3 ± 10.4	4
Lactose, 0.3 M	− 10.6 ± 15.4	6

could not hold, and that the swelling probably involves entrance of solutes as well as water. Wilson (1954) agrees that the process involved here is basically the same as that seen in the colloid-osmotic hemolysis of red cells treated with agents like n-butyl alcohol, and the analogous swelling of lymphocytes in isotonic salt solutions under similar circumstances. The physical permeability of the blood cells to the cations is so low that a moderate degree of surface injury is needed to produce rapid swelling when the ion transport system is inoperative, whereas in the very active tissues the mere cessation of normal ionic transport suffices. In parallel with Deyrup's observations, Wilson found that butyl-alcohol-treated lymphocytes and rat intestine shrink, rather than swell, in disaccharide solutions isosmotic with blood. In view of these findings, one must at least conclude that the condition is more complex than simply an initial hypertonicity of the cytoplasm.

The necessity for maintaining some degree of normal metabolic activity to prevent overhydration *in vitro* has been shown not only for cells or tissue slices, but also more recently for subcellular particulates. Macfarlane and Spencer (1953) found in the rat liver mitochondrial preparations that the failure of the cation accumulation process was accompanied by a considerable water uptake leading to agglutination; DNP did not cause swelling if AMP was present, however, even though no acid-labile P was formed under these circumstances. Price and Davies (1954) extended this study, taking advantage of an optical density change with water content, which permitted rapid continuous estimation of mitochondrial water. They found that the initial water level of about 80% was maintained at 35°C. in the presence of 20 mM succinate and 1 mM ATP or AMP. In the absence of the high-energy phosphate source, the water level rose to 85% in the first half-hour, even though respiratory depression did not appear until later. If the ATP or AMP were added soon enough, this extra water moved back to the medium. An interesting incidental observation was that versene at 3 mM prevented the initial swelling, but could not reverse it.

Water Transport through Intestinal Wall

The movements of water in connection with salt absorption have been followed with D_2O in dog ileal loops *in situ* by VISSCHER, FETCHER *et al.* (1944). Beyond the initial adjustment period, the water migration did not appear to depend particularly on the tonicity of the intestinal contents. At times, the ratio of the two unidirectional fluxes might differ from the ratio of the water "activities" by a factor of 200, so that clearly other factors than diffusion must dictate the passage of water here. Application of $HgCl_2$ at 10^{-3} M completely altered the normal pattern of water movements associated with salt absorption; for details, VISSCHER and ROEPKE (1945) should be consulted.

Water Uptake through Arthropod Surface

Mention should be made of the surprising ability of many insects and arachnids to acquire water directly from fairly humid air. It was suggested by LEES (1947) in his observations on unfed ticks that the water retention in these animals is assisted by an actual water secretion by the epidermal cells, and that this activity is the basis for actual net acquisition of water above the "equilibrium humidity." Killing the ticks led to a marked increase in the rate of water loss even though the cuticle remained intact; a similar observation has been made in spiders (DAVIES and EDNEY 1953).

This topic has been well reviewed by BEAMENT (1954), who presents many interesting data on the movement of water through several types of insect cuticles under various experimental circumstances. This material shows a distinct heterodromy in its water permeability, which is absent after extraction in chloroform. RICHARDS and SCHMITT (1953) studied this property in the isolated cuticles of fly larvae. Although the cuticle shows a perfectly symmetrical water permeability when immersed, the uptake may be 20 times as fast as the loss when water is in contact only with the inner surface, with moist air on the outside. The mechanism of this secretion of water from the vapor into the liquid compartment is not explained.

Water Movements through Amphibian Skin

The osmotic problem of the frog immersed in a very hypotonic pond appears to be worsened by the function of the salt-accumulating system in the skin, discussed earlier, insofar as a certain amount of water seems obliged to accompany the salt taken in by the active process (HUF, PARRISH, and Weatherford 1951; Huf and Wills 1951); thus an additional water load is imposed on the kidney. That this water movement is not simply the result of diffusion along the osmotic gradient across a slightly water-permeable barrier in the skin is demonstrated by unidirectional flux measurements with D_2O (Hevesy, Hofer, and Krogh 1935). The net inward water transfer in the intact frog immersed in water proved to be 3-5 times higher than the figure calculated from the relative water concentrations on the two sides of the skin on the basis of the D_2O diffusion-fluxes through isolated pieces of skin. In other words, the absolute fluxes in the normal asymmetrical situation are not proportional to the water activities of the inner and outer compartments. Koefoed-Johnsen and Ussing (1953) point out that this must mean that an actual solvent flow through pores in the skin is involved [39].

Steggerda (1931) showed that the overhydration of frogs resulting from injection of pitressin did not to a great extent depend on reduced excretion, as the action was still prominent if the cloacas were tied closed in both experimental and control groups. Pitressin did not induce hydration in skinned frogs, nor did it affect the osmotic behavior of the muscles in vitro. It was thus implied that the skin was the site of action. Following Capraro and Bernini's (1952) lead, Koefoed-Johnsen and Ussing (1953) showed that the addition of a posterior pituitary extract did in fact considerably increase the net transport without significant effect on the influx, implying an increase in pore size or opening up of new pores, with only slight increase in the total diffusion area. Adrenaline and Cu^{++}, which affect the Na^+ and Cl^- movements and the p. d. as noted in an earlier section, have little effect on the water flux, but high P_{CO_2} depresses it by about 25%. Sawyer (1951) found the water gain under the action of the posthypophyseal extract to be proportional to the experimentally varied osmotic pressure gradient (Fig. 29), and he concluded that the gradient is the driving force involved. But Capraro (1953) again presents the correlation of skin metabolic activity with the net water transfer, and the responses to hormones, as an argument for "active transport" of the water; this complication seems unnecessary on the basis of Koefoed-Johnsen and Ussing's analysis, but it is clear that the active Na^+ influx does carry some water with it.

Water Transport in Egg Cells

It was shown by Lillie (1916) that fertilization of the egg of *Arbacia* resulted in a quadrupling of the rate at which the egg swelled in dilute sea water; but in both the fertilized and unfertilized egg, the pattern of

---

[39] When an actual bulk flow of the solvent occurs, the classical "permeability constant" determinations lose their intended significance and do not correspond to the flux constants determined by tracer exchange in the steady state; this divergence provides a measure of the porosity of the barrier concerned. An excellent presentation of this argument, and of other complications in the critical interpretation of tracer experiments in connection with cellular permeability, is given by Ussing (1952).

swelling was one of simple diffusion. KEKWIK and HARVEY (1934) observed in the same cell that the swelling in dilute sea water was delayed by the removal of oxygen (washing out with hydrogen), but that the final equilibrium volume was unaltered. This suggested the possibility that the water movements might be in some way associated with the cell's oxidative metabolism.

PRESCOTT and ZEUTHEN (1953) showed that in the eggs of the frog, toad, and several fishes, the filtration permeability constant for water was higher than the diffusion permeability constant, as described above for

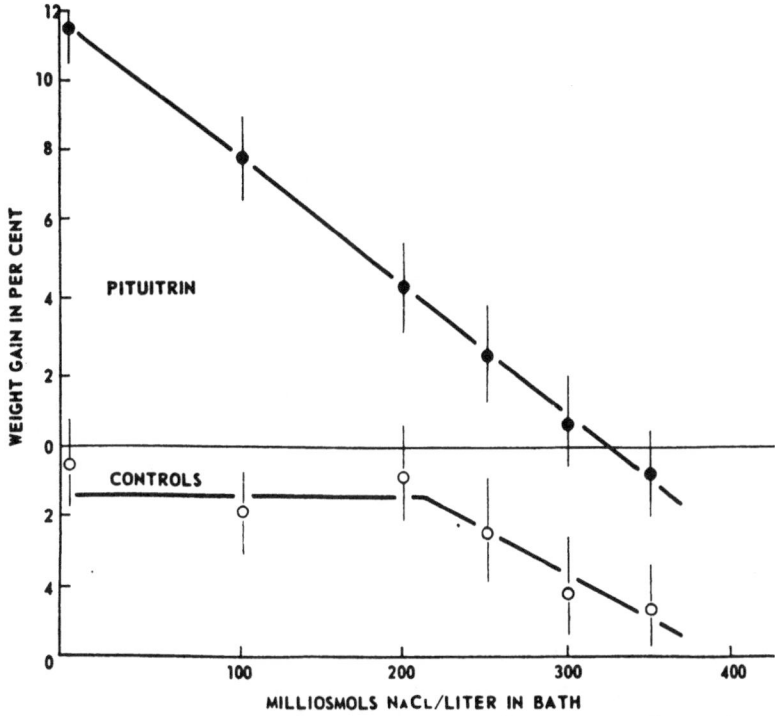

Fig. 29. Abolition by pituitrin of normal frog water-balance response to osmotic concentration of the medium. (Courtesy of W. H. SAWYER and American Journal of Physiology)

other tissues. They made use of D_2O in a special diver balance technique to determine the unidirectional fluxes. Notable differences were tabulated between the various species as regards the numbers of pores and their relative diameters (ratios of the two types of constant varied from 69 for ovarian frog eggs to 1.3 for the shed eggs of the Zebra fish).

Water Transport in the Kidney Tubule

The micro-puncture technique for direct sampling of the renal tubular fluid in the amphibians, mentioned in earlier sections, showed that the specific absorptions which take place in the proximal tubule do not significantly alter the total solute concentrations remaining, so that an essentially isosmotic reabsorption is occurring in this segment. The final dilution is then established in the distal segments. In the mammalian kidney,

7*

Walker et al. (1941) found essentially this same pattern, but here of course the normal final adjustment of tonicity is one of further concentration rather than dilution. In Walker's experiments, the rise in sugar concentration in the proximal tubule after phlorrhizination showed that by far the greater part of the water reabsorption occurred in a presumably passive fashion in this segment, leaving the regulatory adjustment (which would seem to require active processes) to the distal segment.

The same basic arrangement is inferable also by consideration of the urinary alterations produced by disease or experimental upsets. In patients with diabetes insipidus, in which the non-obligatory fraction of the water reabsorption does not proceed normally, Brodsky and Rapoport (1951) found that mannitol loading produced the same type of urinary electrolyte pattern as in the normal individual; thus the disorder is distal at least to the point at which the electrolyte exchanges have already occurred. Similarly, West et al. (1952) showed that the electrolyte patterns produced in a dog's urine by administration of any of a variety of diuretic agents were not altered by extreme variation in the dog's state of hydration. The characteristic electrolyte arrangements in a given part of the filtrate had therefore been achieved before the adjustment had been made of the degree of water reabsorption in that sample, so that the latter must be a distal tubular function. As a result of a direct polarographic cryoscopic study of rat kidney tubule sections, Wirz et al. (1951) placed the site of active water absorption even more distally, in the collecting tubules. According to this study, the whole nephron is an ideally designed hairpin counter-current device which avoids a steep osmotic pressure gradient at any point throughout its entire length: the tonicity is essentially that of the blood in all cortical tubules, and increases in both segments in passing from the cortical-medullary boundary down to the tip of the papilla; thus the fluid is concentrated in descending the loop of Henle, diluted again in ascending the loop, and reconcentrated in the collecting duct.

In spite of the great interest in the active non-obligatory, osmo-regulatory phase of the water transfer in this structure, essentially nothing is known about the cellular mechanisms. The importance therein of an anti-diuretic hormone elaborated from the posterior pituitary, apparently in response to signals from osmoregulatory cells in the supraoptic nucleus in the hypothalamus, has been well established; it is failure of this system which gives rise to the condition of diabetes insipidus in which the water turnover of the body may be enormously increased. However, the nature of the processes in the tubule cells which require the hormone remains mysterious.

Water Extrusion by Contractile Vacuoles

It is generally agreed that the primary function of the contractile vacuoles of the protozoa is the elimination of water accumulated as a result of the osmotic flow of water into the cells from hypotonic surroundings. Most of the marine protozoa do not show these organelles at all, and the activity of the vacuoles in those marine forms in which they do appear is not comparable to that in their brackish or fresh-water relatives. More-over, the activity in any given organism bears an inverse relation to the tonicity of the medium, over the functional range of concentrations. KITCHING (1938 a) has shown for instance (Fig. 30) that dilution of the sea water in which marine forms are immersed leads to an enormous increase

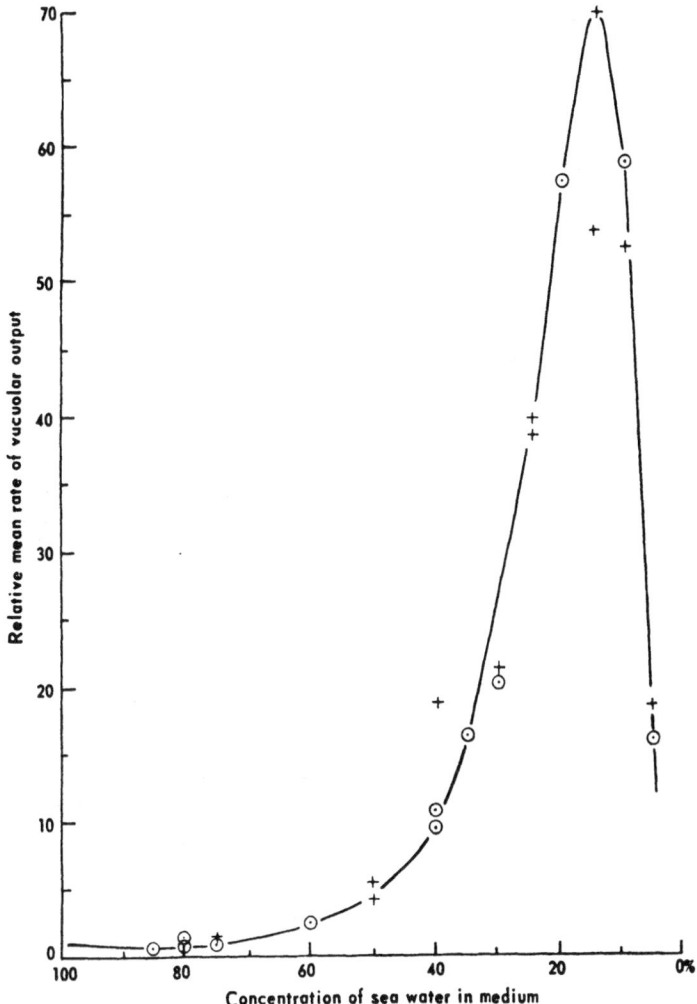

Fig. 30. Relation of rate of vacuolar output to concentration of medium. Circles — *Zoothamnium marinum*; crosses — *Cothuria curvula*.
(Courtesy of Biological Reviews; from KITCHING, 1938 a)

in the vacuolar output; such treatment may even induce the appearance of the vacuole in certain forms in which it is not discernible so long as the cell is in sea water.

Virtually nothing has been learned about the process by which water is apparently extruded into the vacuole during the diastolic phase. KITCHING (1938 b) demonstrated a dependence on metabolic processes, largely in various species of *Zoothamnium*; sulfide or cyanide at fairly weak concentrations greatly diminished the output of the contractile vacuole, and permitted abnormal cell swelling in the absence of sufficient extracellular osmotic pressure (Fig. 31). Sucrose at about 50 mM appeared to be just sufficient to maintain the original cell volume in the presence of cyanide at a level essentially completely stopping the vacuole; thus this was taken as the order of magnitude of the osmotic pressure difference ordinarily maintained.

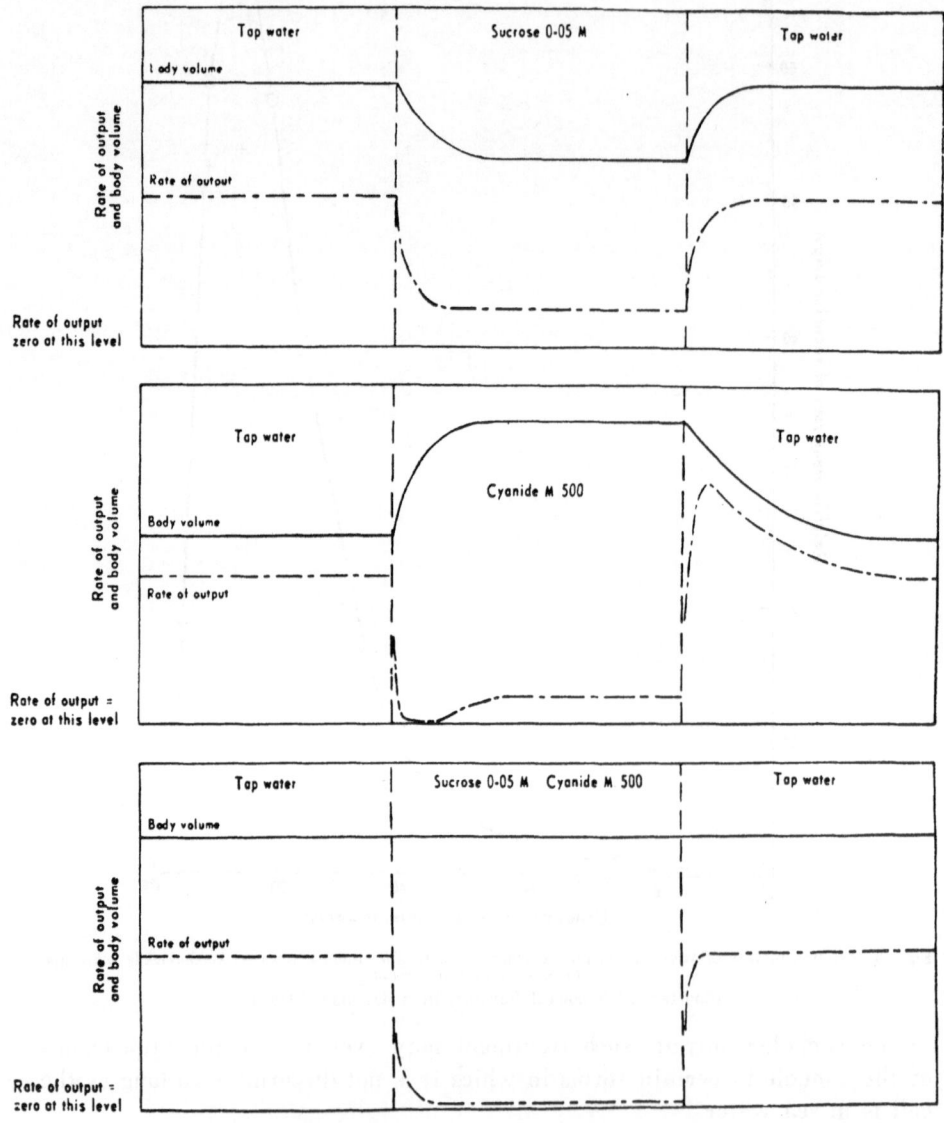

Fig. 31. Diagrammatic illustration of effects of cyanide and sucrose on the body volume and vacuolar output of a fresh-water peritrich (probably a *Zoothamnium*).

(Courtesy of Biological Reviews; from Kitching, 1938 a)

Conclusion

The recurrence throughout this report of discussions of seemingly erratic differences in the operation of even closely related transport systems, in different organisms or in different types of cells, attests to the impossibility, at this stage in the development of our knowledge, of drawing any significant generalizations as to the characteristics of these processes. It would seem reasonable to anticipate that with further study the significance of these differences will become more manifest, and that basic patterns in the structure and functional mechanics of the active transport phenomena will emerge. It is perhaps worthy of comment that relatively few types of cellular membrane-permeation have proved upon critical examination to conform entirely to a passive diffusion pattern, and that new examples of the cell's participation in such processes have been coming to light over the past few decades at an ever increasing pace. The picture gradually developing is that the *bulk* of the *net* material transfer through the membrane or cortical layer of the typical healthy animal cell in the course of the cell's ordinary activities does not result simply from movement along activity gradients through pores or through the substance of the membrane, but is the result of specific translocation processes for which the cell is responsible. The growth of this notion might well be cited as the principal theme of the progress of the past generation in the field of cellular physiology which has until recently passed under the label, "Permeability."

Literature cited

ABDERHALDEN, E., und G. EFFKEMANN, 1934: Über den Einfluß von α- und β-Glucosiden auf die Phosphorylierung von Traubenzucker. Biochem. Z. **268**, 461—468.
— und E. TETZNER, 1935: Beitrag zur Kenntnis des Verhaltens racemischer Aminosäuren im tierischen Organismus. Z. physiol. Chem. **232**, 79—86.
ABELSON, P. H., 1947: Permeability of eggs of *Arbacia punctulata* to radioactive phosphorus. Biol. Bull. (Am.) **93**, 203.
AEBI, H., 1950: Kationenmilieu und Gewebsatmung. Helv. Physiol. Acta **8**, 525—543.
— 1951: Die Bedeutung des Kaliums für die Atmung und Osmoregulation von Leberschnitten. Experientia **7**, 346—347.
— 1953: Elektrolyt-Akkumulierung und Osmoregulation in Gewebschnitten. Helv. Physiol. Pharmacol. Acta **11**, 96—121.
— und A. MEYER, 1951: Das Osmometer-Verhalten von Leberschnitten. Ein Versuch zur Bestimmung des kolloidosmotischen Druckes an isoliertem, überlebendem Gewebe. Helv. Physiol. Acta **9**, C 51—C 52.
ALLAN, F. N., B. R. DICKSON, and J. MARKOWITZ, 1924: The relationship of phosphate and carbohydrate metabolism. II. The effect of adrenalin and phloridzin on the excretion of phosphate. Amer. J. Physiol. **70**, 333—343.
AUBEL, E., et J. SZULMAJSTER, 1950: Contribution à l'étude de la fermentation et de la respiration de *Escherichia coli*. IV. Rôle de la permeabilité dans l'étude du métabolisme bactérien de *E. coli*. Biochim. Biophys. Acta **5**, 515—523.
AUCHINLACHIE, D. W., J. J. R. MACLEOD, and H. E. MAGEE, 1930: Studies on diffusion through surviving isolated intestine. J. Physiol. (Brit.) **69**, 185—209.
BALDWIN, D., E. M. KAHANA, and R. W. CLARKE, 1950: Renal excretion of sodium and potassium in the dog. Amer. J. Physiol. **162**, 655—664.
BANG, O., and S. L. ØRSKOV, 1937: Variations in permeability of red blood cells in man, with particular reference to conditions obtaining in pernicious anemia. J. clin. Invest. (Am.) **16**, 279—288.

Bárány, E., and E. Sperber, 1939: Absorption of glucose against a concentration gradient by the small intestine of the rabbit. Skand. Arch. Physiol. 81, 290—299.

Barnes, R. H., A. N. Wick, E. S. Miller, and E. M. Mackay, 1939: Effect of adrenalectomy on rate of fat absorption. Proc. Soc. exper. Biol. a. Med. (Am.) 40, 651—655.

Barron, E. S. G., J. A. Muntz, and B. Gasvoda, 1948: Regulatory mechanisms of cellular respiration. I. The rôle of cell membranes: uranium inhibition of cellular respiration. J. gen. Physiol. (Am.) 32, 163—178.

Bartlett, G. R., A. N. Wick, and E. M. Mackay, 1949: The influence of insulin and adrenal cortical compounds on the metabolism of radioactive C^{14}-glucose in the isolated rat diaphragm. J. biol. Chem. (Am.) 178, 1003—1004.

Bartley, W., and R. E. Davies, 1952: Secretory activity of mitochondria. Biochem. J. 52, xx—xxi.

— — 1954: Active transport of ions by sub-cellular particles. Biochem. J. 57, 37—49.

— — and H. A. Krebs, 1954: Active transport in animal tissues and subcellular particles. Proc. roy. Soc., Lond. B 142, 187—196.

Bavetta, L., 1943: The effect of adrenalectomy on the absorption of the short chain fatty acids and their triglycerides. Amer. J. Physiol. 140, 44—46.

— L. Hollman, H. J. Deuel Jr., and P. O. Greeley, 1941: The effect of adrenalectomy on fat absorption. Amer. J. Physiol. 134, 619—622.

Beament, J. W. L., 1954: Water transport in insects. Symp. Soc. exper. Biol. 8 (in press).

Beck, L. V., 1942 a: Organic phosphate and "fructose" in rat intestinal mucosa, as affected by glucose and by phlorhizin. J. biol. Chem. (Am.) 143, 403—415.

— 1942 b: Action of phlorhizin on acid phosphatase activity and on glucose phosphorylation of kidney cortex extracts. Proc. Soc. exper. Biol. a. Med. (Am.) 49, 435—439.

Berliner, R. W., and T. J. Kennedy Jr., 1948: Renal tubular secretion of potassium in the normal dog. Proc. Soc. exper. Biol. a. Med. (Am.) 67, 542—545.

— — and J. G. Hilton, 1950: Renal mechanisms for excretion of potassium. Amer. J. Physiol. 162, 348—367.

— — and J. Orloff, 1951: Relationship between acidification of the urine and potassium metabolism. Effect of carbonic anhydrase inhibition on potassium excretion. Amer. J. Med. 11, 274—282.

Beyer, K. H., R. H. Painter, and V. D. Wiebelhaus, 1950: Enzymatic factors in renal tubular secretion of phenol red. Amer. J. Physiol. 161, 259—267.

— L. D. Wright, H. F. Russo, H. R. Skeggs, and E. A. Patch, 1946: The renal clearance of essential amino acids: tryptophane, leucine, isoleucine and valine. Amer. J. Physiol. 146, 330—335.

— — H. R. Skeggs, H. F. Russo, and G. A. Shaner, 1947: Renal clearance of essential amino acids: their competition for reabsorption by the renal tubules. Amer. J. Physiol. 151, 202—210.

Blickenstaff, D., D. M. Bachman, M. E. Steinberg, and W. B. Youmans, 1951: Intestinal absorption of sodium chloride solutions as influenced by intraluminal pressure and concentration. Amer. J. Physiol. 167, 768.

Blowers, R., E. M. Clarkson, and M. Maizels, 1951: Flicker phenomenon in human erythrocytes. J. Physiol. (Brit.) 113, 228—239.

Bogdanove, E. M., and S. B. Barker, 1950: Effect of phlorhizin on intestinal absorption of glucose, galactose, fructose, mannose, and sorbose. Proc. Soc. exper. Biol. a. Med. (Am.) 75, 77—80.

Bornstein, J., and C. R. Park, 1953: Inhibition of glucose uptake by the serum of diabetic rats. J. Biol. Chem. (Am.) 205, 503—511.

Bouckaert, J. P., and C. de Duve, 1947: The action of insulin. Physiol. Rev. (Am.) 27, 39—71.

Boyle, P. J., and E. J. Conway, 1941: Potassium accumulation in muscle and associated changes. J. Physiol. (Brit.) 100, 1—63.

Brodsky, W. A., and S. Rapoport, 1951: The mechanism of polyuria of diabetes insipidus in man. The effect of osmotic loading. J. clin. Invest. (Am.) 30, 282—291.

Brooks, S. C., 1943 a: Intake and loss of ions by living cells. I. Eggs and larvae of *Arbacia punctulata* and *Asterias forbesii* exposed to phosphate and sodium ions. Biol. Bull. (Am.) 84, 213—225.

— 1943 b: Intake and loss of ions by living cells. II. Early changes of phosphate content of *Fundulus* eggs. Biol. Bull. (Am.) 84, 226—239.

BROOKS, S. C. and E. L. CHAMBERS, 1954: The penetration of radioactive phosphate into marine eggs. Biol. Bull. (Anm.) 106, 279—296.

BRÜCKNER, J., 1951: Beeinflussung der selektiven Zuckerresorption durch Phlorrhizin, 2, 4-Dinitrophenol und Atebrin. Helv. Physiol. Acta 9, 259—268.

CALDWELL, P. C., and Sir C. HINSHELWOOD, 1951: The phosphorus metabolism of B. lactis aerogenes. J. Chem. Soc. (1951), 158—166.

CALKINS, E., I. M. TAYLOR, and A. B. HASTINGS, 1954: Potassium exchange in the isolated rat diaphragm; effect of anoxia and cold. Amer. J. Physiol. 117, 211—218.

CAPRARO, V., 1953: Über den aktiven Wassertransport durch die Froschhaut. XIXth Internat. Physiol. Congr., 259—260.

— and G. BERNINI, 1952: Mechanism of action of extracts of the posthypophysis on water transport through the skin of the frog (Rana esculenta). Nature 169, 454.

CARROLL, T. C. N., C. J. DANBY, A. A. EDDY, and Sir C. HINSHELWOOD, 1950: The uptake of alkali metals by bacteria. J. Chem. Soc. (1950), 946—949.

CAUSEY, G., and E. J. HARRIS, 1951: The uptake and loss of phosphate by frog muscle. Biochem. J. 49, 176—183.

CHAMBERS, E. L., and W. E. WHITE, 1949: The accumulation of phosphate and evidence for synthesis of adenosine triphosphate in the fertilized sea-urchin egg. Biol. Bull. (Am.) 97, 225—226.

— — 1954: The accumulation of phosphate by fertilized sea urchin eggs. Biol. Bull. (Am.) 106, 297—307.

— — N. JEUNG, and S. C. BROOKS, 1948: Penetration and effects of low temperature and cyanide on penetration of radioactive potassium into the eggs of Strongylocentrotus purpuratus and Arbacia punctulata. Biol. Bull. (Am.) 95, 252—253.

CHRISTENSEN, H. N., M. K. CUSHING, and J. A. STREICHER, 1949: Concentration of amino-acids by the excised diaphragm suspended in artificial media. II. Inhibition of the concentration of glycine by amino acids and related substances. Arch. Biochem. 23, 106—110.

— and M. E. HENDERSON, 1952: Comparative uptake of free amino acids by mouse-ascites carcinoma cells and normal tissues. Cancer Res. 12, 229—231.

— B. HESS, and T. R. RIGGS, 1954: Concentration of taurine, β-alanine, and triiodothyronine by ascites carcinoma cells. Cancer Res. 14, 124—127.

— and T. R. RIGGS, 1951: Physostigmine uptake by cells and its effect on potassium exchange. J. biol. Chem. (Am.) 193, 621—626.

— — 1952: Concentrative uptake of amino acids by the Ehrlich mouse ascites carcinoma cell. J. biol. Chem. (Am.) 194, 57—68.

— — and B. A. COYNE, 1954: Effects of pyridoxal and indoleacetate on cell uptake of amino acids and potassium. J. biol. Chem. (Am.) 209, 413—427.

— — H. FISCHER, and I. M. PALATINE, 1952 a: Amino acid concentration by a free cell neoplasm: relations among amino acids. J. biol. Chem. (Am.) 198, 1—15.

— — — — 1952 b: Intense concentration of α, γ-diaminobutyric acid by cells. J. biol. Chem. (Am.) 198, 17—22.

— — and N. E. RAY, 1952: Concentrative uptake of amino acids by erythrocytes in vitro. J. biol. Chem. (Am.) 194, 41—51.

— and J. A. STREICHER, 1948: Association between rapid growth and elevated cell concentrations of amino acids. I. In fetal tissues. J. biol. Chem. (Am.) 175, 95—100.

— — 1949: Concentration of amino acids by the excised diaphragm suspended in artificial media. I. Maintenance and inhibition of the concentrating activity. Arch. Biochem. 23, 96—105.

— — and R. L. ELBINGER, 1948: Effects of feeding individual amino acids upon the distribution of other amino acids between cells and extracellular fluid. J. biol. Chem. (Am.) 172, 515—524.

CICARDO, V. H., and J. A. MOGLIA, 1940: Liberation of potassium from muscle by acetylcholine. Nature 145, 551.

CLARK, G. A., 1922: Glucose absorption in the renal tubules of the frog. J. Physiol. (Brit.) 56, 201—205.

CLARKE, E. W., Q. H. GIBSON, D. H. SMYTH, and G. WISEMAN, 1951: Selective absorption of amino-acids from Thiry-Vella loops. J. Physiol. (Brit.) 112, 46 P.

CLARKSON, E. M., and M. MAIZELS, 1954: Respiration, glycolysis and sodium transport in chicken erythrocytes. J. Physiol. (Brit.) 124, 19 P—20 P.

Cohen, P. P., 1939: Microdetermination of glutamic acid. Biochem. J. 33, 551—560.
Collet, R. A., et P. Favarger, 1951: Renouvellement du glycérol dans les phospholipides pendant la résorption intestinale des graisses. Helv. Physiol. Pharmacol. Acta 9, C 61—C 62.
Conway, E. J., 1942: Potassium, fermentation and the cell membrane. Nature 150, 461—462.
— 1946: Ionic permeability of skeletal muscle fibres. Nature 157, 715—717.
— 1951: The biological performance of osmotic work. A redox pump. Science 113, 270—273.
— 1954: Some aspects of ion transport through membranes. Symp. Soc. exper. Biol. 8 (in press).
— M. Carey, and P. T. Moore, 1950: Concerning the entrance rate of KCl into the whole isolated sartorius of the frog and into single fibres. Biochem. J. 47, iii—iv.
— and M. Downey, 1950: An outer metabolic region of the yeast cell. Biochem. J. 47, 347—360.
— O. Fitzgerald, and T. C. MacDougald, 1946: Potassium accumulation in the proximal convoluted tubules of the frog's kidney. J. gen. Physiol. (Am.) 29, 305—334.
— and D. Hingerty, 1948: Relations between potassium and sodium levels in mammalian muscle and blood plasma. Biochem. J. 42, 372—376.
— and J. I. McCormack, 1953: The total intracellular concentration of mammalian tissues compared with that of the extracellular fluid. J. Physiol. (Brit.) 120, 1—14.
— and P. T. Moore, 1950: The azide effect in yeast with respect to potassium and phosphate permeability. Biochem. J. 47, iii.
— and E. O'Malley, 1943: Linkage of physico-chemical processes in biological systems. Nature 151, 252.
— — 1944: Nature of the cation exchanges during short-period yeast fermentation. Nature 153, 555—556.
— — 1946: The nature of the cation exchanges during yeast fermentation, with formation of 0.02 N - H ion. Biochem. J. 40, 59—67.
Cori, C. F., 1925: The fate of sugar in the animal body. I. The rate of absorption of hexoses and pentoses from the intestinal tract. J. biol. Chem. (Am.) 66, 691—715.
— 1926 a: The rate of absorption of a mixture of glucose and galactose. Proc. Soc. exper. Biol. a. Med. (Am.) 23, 290—291.
— 1926 b: The absorption of glycine and d, l-alanine. Proc. Soc. exper. Biol. a. Med. 24, 125—126.
— G. T. Cori, and H. L. Goltz, 1929: On the mechanism of glucose absorption from the intestinal tract. Proc. Soc. exper. Biol. a. Med. (Am.) 26, 433—436.
Cowie, D. B., R. B. Roberts, and I. Z. Roberts, 1949: Potassium metabolism in Escherichia coli. I. Permeability to sodium and potassium ions. J. cellul. a. comp. Physiol. (Am.) 34, 243—258.
Crampton, R. F., Q. H. Gibson, and D. H. Smyth, 1951: The excretion of the D- and L-isomers of amino-acids in the urine. J. Physiol. (Brit.) 115, 7 P.
Creese, R., 1951: Exchangeability of muscle potassium. J. Physiol. (Brit.) 115, 23 P.
— 1952: Bicarbonate ion and muscle potassium. Biochem. J. (Brit.) 50, xviii.
Cross, R. J., and J. V. Taggart, 1950: Renal tubular transport: accumulation of p-aminohippurate by rabbit kidney slices. Amer. J. Physiol. 161, 181—190.
Cumings, J. N., 1940: The rôle of potassium in myasthenia gravis. J. Neur. Psychiat. 3, 115—122.
Danowski, T. S., 1941: The transfer of potassium across the human blood cell membrane. J. biol. Chem. (Am.) 139, 693—705.
Darlington, W. A., and J. H. Quastel, 1953: Absorption of sugars from isolated surviving intestine. Arch. Biochem. Biophys. 43, 194—207.
Davies, M. E., and E. B. Edney, 1953: The evaporation of water from spiders. J. exper. Biol. 29, 571—582.
Davies, R., J. P. Folkes, E. F. Gale, and L. C. Bigger, 1953: The assimilation of amino-acids by micro-organisms. 16. Changes in sodium and potassium accompanying the accumulation of glutamic acid or lysine by bacteria and yeast. Biochem. J. 54, 430—437.
Davies, R. E., and A. W. Galston, 1951: Rapid rate of turnover of potassium ions in kidney slices. Nature 168, 700.

DAVIES, R. E. and H. A. KREBS, 1952: Biochemical aspects of the transport of ions by nervous tissue. Biochem. J. **50**, xxv.

DAVSON, H., 1939: Studies on the permeability of erythrocytes. VI. The effect of reducing the salt content of the medium surrounding the cell. Biochem. J. **33**, 389—401.

— 1940: Ionic permeability. The comparative effects of environmental changes on the permeability of the cat erythrocyte membrane to sodium and potassium. J. cellul. a. comp. Physiol. (Am.) **15**, 317—330.

— 1941: The effect of some metabolic poisons on the permeability of the rabbit erythrocyte to potassium. J. cellul. a. comp. Physiol. (Am.) **18**, 173—185.

— 1951: Textbook of General Physiology, Philadelphia.

— and J. F. DANIELLI, 1938: Studies on the permeability of erythrocytes. V. Factors in cation permeability. Biochem. J. **32**, 991—1001.

— — 1943: The Permeability of Natural Membranes, Cambridge.

— and J. M. REINER, 1942: Ionic permeability; enzyme-like factor concerned in migration of sodium through cat erythrocyte membrane. J. cellul. a. comp. Physiol. (Am.) **20**, 325—342.

DEAN, R. B., 1940: Anaerobic loss of potassium from frog muscle. J. cellul. a. comp. Physiol. (Am.) **15**, 189—193.

— T. R. NOONAN, L. HAEGE, and W. O. FENN, 1941: Permeability of erythrocytes to radioactive potassium. J. gen. Physiol. (Am.) **24**, 353—365.

DEMIS, D. J., 1953: A study of the effects of insulin and of mercury on the utilization of monosaccharides by excised rat diaphragm. University of Rochester Atomic Energy Project, Report UR—297.

DESMEDT, J., 1953: Electrical activity and intracellular sodium concentration in frog muscle. J. Physiol. (Brit.) **121**, 191—205.

DEYRUP, I., 1953: A study of the fluid uptake of rat kidney slices *in vitro*. J. gen. Physiol. (Am.) **36**, 739—749.

DIXON, K. C., 1949: Anaerobic leakage of potassium from brain. Biochem. J. **44**, 187—190.

DONHOFFER, Sz., 1935: Über die elektive Resorption der Zucker. Arch. exper. Path. (D.) **177**, 689—692.

DOTY, J. R., 1941: Reabsorption of certain amino acids and derivatives by the kidney tubules. Proc. Soc. exper. Biol. a. Med. (Am.) **46**. 129—130.

DRURY, D. R., and A. N. WICK, 1951: Insulin and the volume of distribution of glucose. Amer. J. Physiol. **166**, 159—164.

— — 1952: Insulin and cell permeability to galactose. Amer. J. Physiol. **171**, 721.

— — 1953: The nature of the action of insulin. XIXth Internat. Physiol. Congr. 319—320.

EDDY, A. A., T. C. N. CARROLL, C. J. DANBY, and Sir C. HINSHELWOOD, 1951: Alkali-metal ions in the metabolism of *Bact. lactis aerogenes*. I. Experiments on the uptake of radioactive potassium, rubidium and phosphorus. Proc. roy. Soc., Lond. B **138**. 219—228.

— and Sir C. HINSHELWOOD, 1950: The utilization of potassium by *Bact. lactis aerogenes*. Proc. roy. Soc., Lond. B **136**, 544—562.

— — 1951: Alkali-metal ions in the metabolism of *Bact. lactis aerogenes*. III. General discussion of their role and mode of action. Proc. roy. Soc., Lond. B **138**, 237—240.

EGE, R., 1919: Studier over glukosens fordeling mellem plasmaet og de røde blod-legemer. Thesis, Copenhagen. Cited by BANG and ØRSKOV (1937).

— E. GOTTLIEB, and N. W. RAKESTRAW, 1925: The distribution of glucose between human blood plasma and red corpuscles and the rapidity of its penetration. Amer. J. Physiol. **72**, 76—83.

— and K. M. HANSEN, 1927: Distribution of sugar between plasma and red blood corpuscles in man. Acta med. scand. (Schwd.) **65**, 279—299.

EGGLETON, M. G., and Y. A. HABIB, 1950: Urinary excretion of phosphate in man and the cat. J. Physiol. (Brit.) **111**. 423—436.

— and S. SHUSTER, 1954 a: Glucose and phosphate excretion in the cat. J. Physiol. (Brit.) **124**, 613—622.

— — 1954 b: The effect of insulin on the excretion of glucose and phosphate by the kidney of the cat. J. Physiol. (Brit.) **124**, 623—630.

EISENMANN, A. J., L. OTT, P. K. SMITH, and A. W. WINKLER, 1940: Permeability of human erythrocytes to potassium, sodium, and inorganic phosphate by the use of radioactive isotopes. J. biol. Chem. (Am.) **135** 165—173.

Ellinger, P., et A. Lambrechts, 1937: La localisation de l'effet de la phlorhizine dans le rein vivant. C. r. Soc. Biol. 124, 261—263.

Elliott, K. A. C., 1946: Swelling of brain slices and the permeability of brain cells to glucose. Proc. Soc. exper. Biol. a. Med. (Am.) 63, 234—236.

Elsden, S. R., Q. H. Gibson, and G. Wiseman, 1950: Selective absorption of amino-acids from the small intestine of the rat. J. Physiol. (Brit.) 111, 56 P.

Emmens, C. W., and A. W. Blackshaw, 1953: The fertility of ram and bull semen after deep freezing. XIXth Internat. Physiol. Congr. 334—335.

Fenn, W. O., 1937: Loss of potassium in voluntary contraction. Amer. J. Physiol. 120, 675—680.
— and D. M. Cobb, 1936: Electrolyte changes in muscle during activity. Amer. J. Physiol. 115, 345—356.
— — J. F. Manery, and W. R. Bloor, 1937: Electrolyte changes in cat muscle during stimulation. Amer. J. Physiol. 121, 595—608.
— and R. Gerschman, 1950: The loss of potassium from frog nerves in anoxia and other conditions. J. gen. Physiol. (Am.) 33, 195—203.
— R. H. Koenemann, and E. T. Sheridan, 1940: The potassium exchange of per-fused frog muscle during asphyxia. J. cellul. a. comp. Physiol. (Am.) 16, 225—264.
— T. R. Noonan, L. J. Mullins, and L. Haege, 1942: The exchange of radioactive potassium with body potassium. Amer. J. Physiol. 135, 149—163.

Fenton, P. F., 1945: Response of the gastrointestinal tract to ingested glucose solutions. Amer. J. Physiol. 144, 609—619.

Feyder S., and H. B. Pierce, 1935: Rates of absorption and glycogenesis from various sugars. J. Nutrit. (Am.) 9, 435—455.

Fisher, R. B., and D. B. Lindsay, 1954: The effect of insulin on the penetration of galactose into the perfused rat heart. J. Physiol. (Brit.) 124, 20 P—21 P.
— and D. S. Parsons, 1953 a: Glucose movements across the wall of the rat small intestine. J. Physiol. (Brit.) 119, 210—223.
— — 1953 b: Galactose absorption from the surviving small intestine of the rat. J. Physiol. (Brit.) 119, 224—232.

Flemister, L. J., and S. C. Flemister, 1951: Chloride ion regulation and oxygen consumption in the crab Ocypode albicans (Bosq). Biol. Bull. (Am.) 101, 259—273.

Flynn, F., and M. Maizels, 1950: Cation control in human erythrocytes. J. Physiol. (Brit.) 110, 301—318.

Forster, R. P., and J. V. Taggart, 1950: Use of isolated renal tubules for the examination of metabolic processes associated with active cellular transport. J. cellul. a. comp. Physiol. (Am.) 36, 251—270.

Foulks, J., P. Brazeau, E. S. Koelle, and A. Gilman, 1952: Renal secretion of thiosulfate in the dog. Amer. J. Physiol. 168, 77—85.
— G. H. Mudge, and A. Gilman, 1952: Renal excretion of cation in the dog during infusion of isotonic solutions of lithium chloride. Amer. J. Physiol. 168, 642—649.

Franck, J., and J. E. Mayer, 1947: An osmotic diffusion pump. Arch. Biochem. 14, 297—313.

Fridlander, L., and J. H. Quastel, 1953: Absorption of sugars and amino acids by isolated surviving intestine. XIXth Internat. Physiol. Congr. 365—366.

Fuhrman, F. A., 1952: Inhibition of active sodium transport in the isolated frog skin. Amer. J. Physiol. 171, 266—278.
— and H. H. Ussing, 1951: A characteristic response of the isolated frog skin potential to neurohypophysial principles and its relation to the transport of sodium and water. J. cellul. a. comp. Physiol. (Am.) 38, 109—130.

Furchgott, R. F., and E. Shorr, 1943: Phosphate exchange in resting cardiac muscle as indicated by radioactivity studies. IV. J. biol. Chem. (Am.) 151, 65—86.

Gale, E. F., 1947 a: The assimilation of amino-acids by bacteria. I. The passage of certain amino-acids across the cell wall and their concentration in the internal environment of Streptococcus faecalis. J. gen. Microbiol. 1, 53—76.
— 1947 b: The assimilation of amino-acids by bacteria. 6. The effect of protein synthesis on glutamic acid accumulation and the action thereon of sulpha-thiazole. J. gen. Microbiol. 1, 327—334.

GALE, E. F.. 1949: The assimilation of amino-acids by bacteria. 8. Trace metals in glutamic acid assimilation and their inactivation by 8-hydroxyquinoline. J. gen. Microbiol. 3, 369—386.

— 1951 a: The assimilation of amino-acids by bacteria. 10. Action of inhibitors on the accumulation of free glutamic acid in *Staphylococcus aureus* and *Streptococcus faecalis*. Biochem. J. 48, 286—290.

— 1951 b: The assimilation of amino-acids by bacteria. 11. The relationship between accumulation of free glutamic acid and the formation of combined glutamic acid in *Staphylococcus aureus*. Biochem. J. 48, 290—297.

— 1953: Assimilation of amino-acids by gram-positive bacteria and some actions of antibiotics thereon. Adv. in Prot. Chem. 8, 285—391.

— 1954: The accumulation of amino-acids within staphylococcal cells. Symp. Soc. exper. Biol. 8 (in press).

— and P. D. MITCHELL, 1947: The assimilation of amino-acids by bacteria. 4. The action of triphenylmethane dyes on glutamic acid assimilation. J. gen. Microbiol. 1, 299—313.

— and A. W. RODWELL, 1948: Amino acid metabolism of penicillin-resistant staphylococci. J. Bacter. (Am.) 55, 161—167.

— — 1949: The assimilation of amino-acids by bacteria. 7. The nature of resistance to penicillin in *Staphylococcus aureus*. J. gen. Microbiol. 3, 127—142.

— and E. S. TAYLOR, 1947: The assimilation of amino-acids by bacteria. 5. The action of penicillin in preventing the assimilation of glutamic acid by *Staphylococcus aureus*. J. gen. Microbiol. 1, 314—326.

GAMMELTOFT, A., and K. KJERULF-JENSEN, 1943: The mechanism of renal excretion of fructose and galactose in rabbit, cat, dog and man (with special reference to the phosphorylation theory). Acta Physiol. Scand. (D.) 6, 368—384.

GEIGER, A., J. MAGNES, and J. DOBKIN, 1933: The role of a liver factor in maintaining the glucose uptake, carbohydrate metabolism and the responsiveness of the brain. The utilization of glucosamine. XIXth Internat. Physiol. Congr. 383—384.

— — R. M. TAYLOR, and M. VERALLI. 1954: Effect of blood constituents on uptake of glucose and on metabolic rate of the brain in perfusion experiments. Amer. J. Physiol. 177, 138—149.

GEMMILL, C. L., and L. HAMMAN Jr., 1941: The effect of insulin on glycogen deposition and on glucose utilization by isolated muscles. Bull. Johns Hopkins Hosp. 68, 50—57.

GIBSON, Q. H., and G. WISEMAN, 1951: Selective absorption of stereo-isomers of amino-acids from loops of the small intestine of the rat. Biochem. J. 48, 426—429.

GOLDSTEIN, M. S., W. L. HENRY, B. HUDDLESTUN. and R. LEVINE, 1953: Action of insulin on transfer of sugars across cell barriers: common chemical configuration of substances responsive to action of the hormone. Amer. J. Physiol. 173, 207—211.

— V. MULLICK, B. HUDDLESTUN, and R. LEVINE. 1953: Action of muscular work on transfer of sugars across cell barriers: comparison with action of insulin. Amer. J. Physiol. 173, 212—216.

GOURLEY, D. R. H., 1951: Inhibition of uptake of radioactive phosphate by human erythrocytes *in vitro*. Amer. J. Physiol. 164, 213—220.

— 1952: The role of adenosine triphosphate in the transport of phosphate in the human erythrocyte. Arch. Biochem. 40, 1—12.

— and C. L. GEMMILL, 1950: The effect of temperature upon the uptake of radioactive phosphate by human erythrocytes *in vitro*. J. cellul. a. comp. Physiol. (Am.) 35, 341—352.

GREIG, M. E., J. S. FAULKNER, and T. C. MAYBERRY, 1953: Studies on permeability. IX. Replacement of potassium in erythrocytes during cholinesterase activity. Arch. Biochem. Biophys. 43, 39—47.

— and W. C. HOLLAND, 1949 a: Effect of the D- and L-isomers of isoamidone on the permeability of dog erythrocytes. Proc. Soc. exper. Biol. a. Med. (Am.) 71, 189—192.

— — 1949 b: Studies on the permeability of erythrocytes. I. The relationship between cholinesterase activity and permeability of dog erythrocytes. Arch. Biochem. 23. 370—384.

Greig, M. E., and W. C. Holland, 1951: Studies on the permeability of erythrocytes. IV. Effect of certain choline and non-choline esters on permeability of dog erythrocytes. Amer. J. Physiol. 164, 423—427.
— T. C. Mayberry, and C. E. Dunn, 1951: Replacement of potassium in the human erythrocyte during cholinesterase activity. Fed. Proc. 10, 302—303.
Groen, J., 1937: Absorption of hexoses from upper part of small intestine in man. J. clin. Invest. (Am.) 16, 245—255.
Grossfeld, H. D., 1951: Cell permeability to electrolytes in tissue culture. Exper. Cell Res. 2, 141—143.
Grundfest, H., and D. Nachmansohn, 1950: Increased sodium entry into squid giant axons during activity at high frequencies and during reversible inactivation of cholinesterase. Fed. Proc. 9, 53.
Guensberg, E., 1947: Die Glukoseaufnahme in menschliche rote Blutkörperchen. Inauguraldissertation, Bern; Schwarzenburg.
Hahn, L., and G. Hevesy, 1942: Rate of penetration of ions into erythrocytes. Acta Physiol. Scand. (D.) 3, 193—223.
— — and O. H. Rebbe, 1939: Do the potassium ions inside the muscle cells and blood corpuscles exchange with those present in the plasma? Biochem. J. 33, 1549—1558.
Hald, P. M., A. J. Heinsen, and J. P. Peters, 1948: Effects of isotonic solutions and of sulfates and phosphates on the distribution of water and electrolytes in human blood. Amer. J. Physiol. 152, 77—85.
— M. Tulin, T. S. Danowski, P. H. Lavietes, and J. P. Peters, 1947: The distribution of sodium and potassium in oxygenated human blood and their effects upon the movements of water between cells and plasma. Amer. J. Physiol. 149, 340—349.
Halpern, L., 1936: The transfer of inorganic phosphorus across the red blood cell membrane. J. biol. Chem. (Am.) 114, 747—770.
van Harreveld, A., 1950: The potassium permeability of the myelin sheath of a vertebrate nerve. J. cellul. a. comp. Physiol. (Am.) 35, 331—340.
Harris, E. J., 1953 a: The exchange of frog muscle potassium. J. Physiol. (Brit). 120, 246—253.
— 1953 b: Phosphate liberation from isolated frog muscle. J. Physiol. (Brit.) 122, 366—370.
— 1954: Linkage of Na and K transport in human erythrocytes. Symp. Soc. exper. Biol. 8 (in press).
— and G. P. Burn, 1949: The transfer of sodium and potassium ions between muscle and the surrounding medium. Trans. Farad. Soc. 45, 508—528.
— and M. Maizels, 1951: The permeability of human erythrocytes to sodium. J. Physiol. (Brit.) 113, 506—524.
— — 1952: Distribution of ions in suspensions of human erythrocytes. J. Physiol. (Brit.) 118, 40—53.
Harris, J. E., 1941: The influence of the metabolism of human erythrocytes on their potassium content. J. biol. Chem. (Am.) 141, 579—595.
Heinsen, A. J., 1948: Effect of inorganic phosphate on the glycolysis of human blood. Amer. J. Physiol. 152, 216—218.
Heppel, L. A., 1939: The electrolytes of muscle and liver in potassium-depleted rats. Amer. J. Physiol. 127, 385—392.
— 1940: The diffusion of radioactive sodium into the muscles of potassium-deprived rats. Amer. J. Physiol. 128, 449—454.
Hestrin-Lerner, S., and B. Shapiro 1953: Active absorption of glucose from the intestine. Nature 171, 745—746.
Hetényi, G., and M. Winter, 1952: Contributions to the mechanism of the intestinal absorption of amino acids. Acta Physiol. Acad. Sci. Hungar. 3, 49—58.
Hevesy, G., E. Hofer, and A. Krogh, 1935: The permeability of the skin of frogs to water as determined by D_2O and H_2O. Skand. Arch. Physiol. (D.) 72, 199—214.
— and N. Nielsen, 1941: Potassium interchange in yeast cells. Acta Physiol. Scand. 2, 347—354.
Hewitt, J. A., 1924: The metabolism of carbohydrates. Part III. The absorption of glucose, fructose and galactose from the small intestine. Biochem. J. 18, 160—170.

HODGKIN, A. L., 1949: Ionic exchange and electrical activity in nerve and muscle. Arch. Sci. Physiol. 3, 151—163.
— and A. F. HUXLEY, 1952 a: Currents carried by sodium and potassium ions through the membrane of the giant axon of *Loligo*. J. Physiol. (Brit.) 116, 449—472.
— — 1952 b: The components of membrane conductance in the giant axon of *Loligo*. J. Physiol. (Brit.) 116, 473—496.
— — 1952 c: The dual effect of membrane potential on sodium conductance in the giant axon of *Loligo*. J. Physiol. (Brit.) 116, 497—506.
— — 1953: Movement of radioactive potassium and membrane current in a giant axon. J. Physiol. (Brit.) 121, 403—414.
— and R. D. KEYNES, 1953: The mobility and diffusion coefficient of potassium in giant axons from *Sepia*. J. Physiol. (Brit.) 119, 513—528.
— — 1954: Movement of cations during recovery in nerve. Symp. Soc. exper. Biol. 8 (in press).
HÖBER, R., 1945: Physical Chemistry of Cells and Tissues, Philadelphia.
— and J. HÖBER, 1937: Experiments on the absorption of organic solutes in the small intestine of rats. J. cellul. a. comp. Physiol. (Am.) 10, 401—422.
HOGBEN, C. A. M., and J. L. BOLLMAN, 1951: Excretion of phosphate by isolated frog kidney: an "adsorption semipermeability" model for maximal tubular transport. Amer. J. Physiol. 164, 662—669.
HOLLAND, W. C., C. E. DUNN, and M. E. GREIG, 1952 a: Studies on permeability. VII. Effect of several substrates and inhibitors of acetyl cholinesterase on permeability of isolated auricles to Na and K. Amer. J. Physiol. 168, 546—556.
— — — 1952 b: Studies on permeability. VIII. Role of acetylcholine metabolism in the genesis of the electrocardiogram. Amer. J. Physiol. 170, 339—345.
— and M. E. GREIG, 1950 a: Studies on permeability. II. The effect of acetylcholine and physostigmine on the permeability to potassium of dog erythrocytes. Arch. Biochem. 26, 151—155.
— — 1950 b: Studies on the permeability of erythrocytes. III. The effect of physostigmine and acetyl choline on the permeability of dog, cat and rabbit erythrocytes to sodium and potassium. Amer. J. Physiol. 162, 610—615.
— — 1951: Studies on permeability. VI. Increased permeability of dog erythrocytes caused by cholinesterase inhibitors. Arch. Biochem. Biophys. 32, 428—435.
HORVÁTH, I., and G. WIX, 1951: Hormonal influences on glucose resorption from the intestines. I. Methodical principles. Daily variations in the absorption of sugar. The proportion between the absorption of glucose and xylose. Acta Physiol. Acad. Sci. Hungar. 2, 435—443.
HOTCHKISS, R. D., 1944: Gramicidin, tyrocidine, and tyrothricin. Adv. in Enzymol. 4, 153—199.
HUF, E. G., 1935 a: Versuche über den Zusammenhang zwischen Stoffwechsel, Potentialbildung und Funktion der Froschhaut. Arch. ges. Physiol. 235, 655—673.
— 1935 b: Über den Anteil vitaler Kräfte bei der Resorption von Flüssigkeit durch die Froschhaut. Arch. ges. Physiol. 236, 1—19.
— 1936 a: Über aktiven Wasser- und Salztransport durch die Froschhaut. Arch. ges. Physiol. 237, 143—166.
— 1936 b: Die Bedeutung der Atmungsvorgänge für die Resorptionsleistung und Potentialbildung bei der Froschhaut. Biochem. Z. 288, 116—122.
— 1936 c: Die Reproduzierbarkeit des Reidschen Versuchs. Arch. ges. Physiol. 238, 97—102.
— and J. PARRISH, 1951: Nature of the electrolyte pump in surviving frog skin. Amer. J. Physiol. 164, 428—436.
— — and C. WEATHERFORD, 1951: Active salt and water uptake by isolated frog skin. Amer. J. Physiol. 164, 137—142.
— and J. WILLS, 1951: Influence of some inorganic cations on active salt and water uptake by isolated frog skin. Amer. J. Physiol. 167, 255—260.
— — 1953: The relationship of sodium uptake, potassium rejection, and skin potential in isolated frog skin. J. gen. Physiol. (Am.) 36, 473—487.
— — and M. J. COOLEY, 1951: The significance of the anion in active salt uptake by isolated frog skin. Arch. ges. Physiol. 255, 16—26.
HUNTER, F. R., 1936: The effect of lack of oxygen on cell permeability. J. cellul. a. comp. Physiol. (Am.) 9, 15—27.
— 1937: Effect of prolonged exposures to lack of oxygen on permeability of erythrocyte. J. cellul. a. comp. Physiol. (Am.) 10, 241—245.

HUNTER, F. R., 1941: Metabolism and permeability. Anat. Rec. (Am.) **81** Suppl., 31—32.
— 1947 a: Further studies on the relationship between cell permeability and metabolism. The effect of certain respiratory inhibitors on the permeability of erythrocytes to non-electrolytes. J. cellul. a. comp. Physiol. (Am.) **29**, 301—312.
— 1947 b: The effect of washing on the permeability and metabolism of chicken erythrocytes. J. cellul. a. comp. Physiol. (Am.) **29**, 313—321.
— and V. PAHIGIAN, 1940: The effect of temperature on cell permeability and on cell respiration. J. cellul. a. comp. Physiol. (Am.) **15**, 387—394.
HURWITZ, L., and A. ROTHSTEIN, 1951: The relationship of the cell surface to metabolism. VII. The kinetics and temperature characteristics of uranium-inhibition. J. cellul. a. comp. Physiol. (Am.) **38**, 437—450.
INGRAHAM, R. C., and M. B. VISSCHER, 1936 a: The production of chloride-free solutions by the action of the intestinal epithelium. Amer. J. Physiol. **114**, 676—680.
— — 1936 b: The influence of various poisons on the movement of chloride against concentration gradients from intestine to plasma. Amer. J. Physiol. **114**, 681—687.
— — 1938: Further studies on intestinal absorption with the performance of osmotic work. Amer. J. Physiol. **121**, 771—785.
JACOBS, M. H., 1931: The permeability of the erythrocyte. Erg. Biol. **7**, 1—55.
— 1950: Surface properties of the erythrocyte. Ann. N. Y. Ac. Sci. **50**, 824—834.
— and S. A. CORSON, 1934: The influence of minute traces of copper on certain hemolytic processes. Biol. Bull. (Am.) **67**, 325—326.
— H. N. GLASSMAN, and A. K. PARPART, 1935: Osmotic properties of the erythrocyte. VII. The temperature coefficients of certain hemolytic processes. J. cellul. a. comp. Physiol. (Am.) **7**, 197—225.
— — 1938: Osmotic properties of the erythrocyte. XI. Differences in the permeability of the erythrocytes of two closely related species. J. cellul. a. comp. Physiol. (Am.) **11**, 479—494.
— and A. K. PARPART, 1933: Osmotic properties of the erythrocyte. VI. The influence of the escape of salts on hemolysis by hypotonic solutions. Biol. Bull. (Am.) **65**, 512—528.
— — 1937: The influence of certain alcohols on the permeability of the erythrocyte. Biol. Bull. (Am.) **73**, 380—381.
— and D. R. STEWART, 1946: Observations on an oligodynamic action of copper on human erythrocytes. Amer. J. med. Sci. **211**, 246.
JØRGENSEN, C. B., 1947: The effect of adrenaline and related compounds on the permeability of isolated frog skin to ions. Acta Physiol. Scand. **14**, 213—219.
— H. LEVI, and H. H. USSING, 1947: On the influence of the neurohypophyseal principles on the sodium metabolism in the axolotl (*Amblystoma mexicanum*). Acta Physiol. Scand. **12**, 350—371.
JOHNSON, C. A., and O. BERGEIM, 1951: The distribution of free amino-acids between erythrocytes and plasma in man. J. Biol. Chem. (Am.) **188**, 833—838.
JONAS, H., 1954: Observations on the mechanism of phosphate uptake by rabbit erythrocytes. Phosphate adsorption in relation to cell surface structure; equilibria of phosphate adsorption and absorption. Biochim. Biophys. Acta **13**, 241—250.
— and D. R. H. GOURLEY, 1954: Effect of adenosine triphosphate, magnesium and calcium on the phosphate uptake by rabbit erythrocytes. Biochim. Biophys. Acta (in press).
JONES, L. L., 1941: Osmotic regulation in several crabs of the Pacific coast of North America. J. cellul. a. comp. Physiol. (Am.) **18**, 79—92.
KABAT, E. A., and J. FURTH, 1941: A histochemical study of the distribution of alkaline phosphatase in various normal and neoplastic tissues. Amer. J. Path. **17**, 303—318.
KAMEN, M. D., and S. SPIEGELMAN, 1948: Studies on the phosphate metabolism of some unicellular organisms. Cold Spring Harbor Symp. Quant. Biol. **13**, 151—163.
KAMIN, H., and P. HANDLER, 1951: Effect of infusion of single amino acids upon excretion of other amino acids. Amer. J. Physiol. **164**, 654—661.
KATZIN, L., I., 1940: The use of radioactive tracers in the determination of irreciprocal permeability of biological membranes. Biol. Bull. (Am.) **79**, 342.
KEKWIK, R. A., and E. N. HARVEY, 1934: The effect of anaerobic conditions on the permeability of the egg of *Arbacia punctulata* to water. J. cellul. a. comp. Physiol. (Am.) **5**, 43—51.

KEYNES, R. D.. 1949: Movements of radioactive ions in resting and stimulated nerve. Arch. Sci. Physiol. 3, 165—175.
— 1951 a: The leakage of radioactive potassium from stimulated nerve. J. Physiol. (Brit.) 113, 99—114.
— 1951 b: The ionic movements during nervous activity. J. Physiol. (Brit.) 114, 119—150.
— 1954: The ionic fluxes in frog muscle. Proc. roy. Soc., Lond. B 142, 359—382.
— and P. R. LEWIS, 1950: Determination of the ionic exchange during nervous activity by activation analysis. Nature 165, 809—810.
— — 1951: The resting exchange of radioactive potassium in crab nerve. J. Physiol. (Brit.) 113, 73—98.
— and G. W. MAISEL, 1954: The energy requirement for sodium extrusion from a frog muscle. Proc. roy. Soc., Lond. B 142, 383—392.
KEYS, A. B.. 1931: Chloride and water secretion and absorption by the gills of the eel. Z. vergl. Physiol. 15. 364—388.
— and E. N. WILLMER, 1932: "Chloride secreting cells" in the gills of fishes, with special reference to the common eel. J. Physiol. (Brit.) 76, 368—378.
KIRSCHNER, L. B., 1953: Effect of cholinesterase inhibitors and atropine on active sodium transport across frog skin. Nature 172, 348—350.
KITCHING, J. A., 1938 a: Contractile vacuoles. Biol. Rev. 13. 403—444.
— 1938 b: The physiology of contractile vacuoles. III. The water balance of fresh-water peritricha. J. exper. Biol. 15, 143—151.
KJERULF-JENSEN, K., und E. LUNDSGAARD, 1940: Quantitative Wertung des Umsatzes der Phosphatester in der Darmschleimhaut von Ratten während der Fructose-resorption. Z. physiol. Chem. 266, 217—224.
KLINGHOFFER, K. A., 1935: Permeability of the red cell membrane to glucose. Amer. J. Physiol. 111, 231—242.
— 1938: The effect of monoiodoacetatic acid on the intestinal absorption of monosaccharides and sodium chloride. J. biol. Chem. (Am.) 126, 201—205.
KLINGMULLER, V. O. G., 1953: Asymmetric absorption, distribution and excretion of optical antipodes. XIXth Internat. Physiol. Congr. 925—926.
KOCH, H. J., 1938: The absorption of chloride ions by the anal papillae of Diptera larvae. J. exper. Biol. 15, 152—160.
KOEFOED-JOHNSEN, V., H. LEVI, and H. H. USSING, 1952: The mode of passage of chloride ions through the isolated frog skin. Acta Physiol. Scand. 25, 150—163.
— and H. H. USSING, 1949: The influence of the corticotropic hormone from ox on the active salt uptake in the axolotl. Acta Physiol. Scand. 17, 38—43.
— — 1953: The contributions of diffusion and flow to the passage of D_2O through living membranes. Effect of neurohypophyseal hormone on isolated anuran skin. Acta Physiol. Scand. 28, 60—76.
— — and K. ZERAHN, 1952: The origin of the short-circuit current in the adrenaline stimulated frog-skin. Acta Physiol. Scand. 27. 38—48.
KOREY, S. R., 1950: Permeability of axonal surface membranes to amino acids. Fed. Proc. 9, 191—192.
— 1952: Studies on permeability in relation to nerve function. IV. Effect of glutamate and aspartate upon the rate of entrance of potassium into brain cortical slices. Biochim. Biophys. Acta 9. 633—635.
— and R. MITCHELL, 1951: Studies on permeability in relation to nerve function. III. Permittivity of brain cortex slices to glycin and aspartic acid. Biochim. Biophys. Acta 7, 507—519.
KOZAWA, S.. 1914: Beiträge zum arteigenen Verhalten der roten Blutkörperchen. III. Artdifferenzen in der Durchlässigkeit der roten Blutkörperchen. Biochem. Z. 60, 231—256.
KRAHL, M. E., 1951: The effect of insulin and pituitary hormones on glucose uptake in muscle. Ann. N.Y. Ac. Sci. 54, 649—670.
— and C. F. CORI, 1947: The uptake of glucose by the isolated diaphragm of normal, diabetic and adrenalectomized rats. J. biol. Chem. (Am.) 170, 607—618.
— and C. R. PARK, 1948: The uptake of glucose by the isolated diaphragm of normal and hypophysectomized rats. J. biol. Chem. (Am.) 174, 939—946.
KREBS, H. A., L. V. EGGLESTON, and C. TERNER, 1951: In vitro measurements of the turnover rate of potassium in brain and retina. Biochem. J. 48, 530—537.
KRITZLER, R. A., and A. B. GUTMAN, 1941: "Alkaline" phosphatase activity of the proximal convoluted tubules and the mechanism of phlorizin glycuresis. Amer. J. Physiol. 134, 94—101.

KROGH, A., 1937 a: Osmotic regulation in the frog (*Rana esculenta*) by active absorption of chloride ions. Skand. Arch. Physiol. **76**, 60—73.
— 1937 b: Active absorption of anions in the animal kingdom. Nature **139**, 755.
— 1937 c: Osmotic regulation in fresh water fishes by active absorption of chloride ions. Z. vergl. Physiol. **24**, 656—666.
— 1938: The active absorption of ions in some freshwater animals. Z. vergl. Physiol. **25**, 335—350.
— 1943: The exchange of ions between cells and extracellular fluid. I. The uptake of potassium into the chorion membrane from the hen's egg. Acta Physiol. Scand. **6**, 203—221.
— 1946: The active and passive exchanges of inorganic ions through the surfaces of living cells and through living membranes generally. Proc. roy. Soc., Lond. B **133**, 140—200.
LAMBRECHTS, A., 1934: Appréciation de la quantité de phlorhizine dans le foie et les reins après injection intraveineuse chez le chien. C. r. Soc. Biol. **116**, 355—357.
— 1936 a: Processus de déphosphorylation pendant le diabète phlorhizique chez le chien. C. r. Soc. Biol. **122**, 72—73.
— 1936 b: Phlorhizine et excrétion urinaire de phosphore. C. r. Soc. Biol. **122**. 468—470.
— 1936: Influence de la phlorhizine sur la phosphatase rénale *in vitro*. C. r. Soc. Biol. **123**, 311—313.
— 1937: Nouvelles recherches sur le diabète phlorhizique, la phlorhizine et quelques substances apparentées. Quatrième memoire: Quelques recherches sur le mécanisme de la glycosurie phlorhizique. Arch. internat. Physiol. **44**, Suppl., 136—162.
LASZT, L., 1935: Die Resorption von Glukose und Xylose bei verschiedener H-Konzentration. Biochem. Z. **276**, 40—43.
— und H. SÜLLMANN, 1935: Nachweis der Bildung von Phosphorsäureestern in der Darmschleimhaut bei der Resorption von Zuckern und Glyzerin. Biochem. Z. **278**, 401—417.
LEES, A. D., 1947: Transpiration and the structure of the epicuticle in ticks. J. exper. Biol. **23**, 379—410.
LEFEVRE, P. G., 1947: Evidence of active transfer across the human erythrocyte membrane. Biol. Bull. (Am.) **93**, 224.
— 1948: Evidence of active transfer of certain non-electrolytes across the human red cell membrane. J. gen. Physiol. (Am.) **31**, 505—527.
— 1953: Further characterization of the sugar-transfer system in the red cell membrane by the use of phloretin. Fed. Proc. **12**, 84.
— 1954: The evidence for active transport of monosaccharides across the red cell membrane. Symp. Soc. exper. Biol. **8** (in press).
— and R. I. DAVIES, 1951: Active transport into the human erythrocyte: evidence from comparative kinetics and competition among monosaccharides. J. gen. Physiol. (Am.) **34**, 515—524.
— and M. E. LEFEVRE, 1952: The mechanism of glucose transfer into and out of the human red cell. J. gen. Physiol. (Am.) **35**, 891—906.
LEIBOWITZ, J., and N. KUPERMINTZ, 1942: Potassium in bacterial fermentation. Nature **150**, 233.
LEVI, H., and H. H. USSING, 1948: The exchange of sodium and chloride ions across the fibre membrane of the isolated frog sartorius. Acta Physiol. Scand. **16**, 232—249.
— — 1949: Resting potential and ion movements in the frog skin. Nature **164**, 928—929.
LEVINE, R., and M. S. GOLDSTEIN, 1953: The effect of insulin on the transfer of sugars across cell barriers. XIXth Internat. Physiol. Congr. 557—558.
— — B. HUDDLESTUN, and S. P. KLEIN, 1950: Action of insulin on the "permeability" of cells to free hexoses, as studied by its effect on the distribution of galactose. Amer. J. Physiol. **163**, 70—76.
— — S. KLEIN, and B. HUDDLESTUN, 1949: The action of insulin on the distribution of galactose in eviscerated nephrectomized dogs. J. biol. Chem. (Am.) **179**, 985—986.
LEVINSKY, N. G., and W. H. SAWYER, 1953: Relation of metabolism of frog skin to cellular integrity and electrolyte transfer. J. gen. Physiol. (Am.) **36**, 607—615.

LILLIE, R. S., 1916: Increase of permeability to water following normal and artificial activation in sea-urchin eggs. Amer. J. Physiol. **40**, 249—266.

LINDBERG, O., 1950: On surface reactions in the sea urchin egg. Exper. Cell. Res. **1**, 105—114.

LINDERHOLM, H., 1952: Active transport of ions through frog skin with special reference to the action of certain diuretics. A study of the relation between electrical properties, the flux of labelled ions, and respiration. Acta Physiol. Scand. **27**, suppl. 97.

— 1953: The electrical potential across isolated frog skins and its dependence on the permeability of the skins to chloride ions. Acta Physiol. Scand. **28**, 211—217.

LINDVIG, P. E., M. E. GREIG, and S. W. PETERSON, 1951: Studies on permeability. V. The effects of acetylcholine and physostigmine on the permeability of human erythrocytes to sodium and potassium. Arch. Biochem. **30**, 241—250.

LING, G. N., 1953: Selective cellular permeability according to the fixed charge hypothesis (FCH). XIXth Internat. Physiol. Congr. 566—567.

LOTSPEICH, W. D., 1947: Renal tubular reabsorption of inorganic sulphate in the normal dog. Amer. J. Physiol. **151**, 311—318.

— R. C. SWAN, and R. F. PITTS, 1947: The renal tubular reabsorption of chloride. Amer. J. Physiol. **148**, 445—448.

LOURAU, M., et O. LARTIGUE, 1951: L'absorption intestinale du glucose chez les cobayes irradiés. Arch. Sci. Physiol. **5**, 83—92.

LUNDSGAARD, E., 1933 a: Hemmung von Esterifizierungsvorgängen als Ursache der Phlorrhizinwirkung. Biochem. Z. **264**, 209—220.

— 1933 b: Die Wirkung von Phlorrhizin auf die Glukoseresorption. Biochem. Z. **264**, 221—223.

— 1935: The effect of phloridzin on the isolated kidey and isolated liver. Skand. Arch. Physiol. **72**, 265—270.

— 1939: Die säurelöslichen Phosphatverbindungen in der Darmschleimhaut bei Ruhe und während der Hexoseresorption. Z. physiol. Chem. **261**, 193—208.

MACFARLANE, M. G., and A. G. SPENCER, 1953: Changes in the water, sodium and potassium content of rat-liver mitochondria during metabolism. Biochem. J. **54**, 569—575.

MACKAY, E. M., and H. C. BERGMAN, 1933: The rate of absorption of glucose from the intestinal tract. J. biol. Chem. (Am.) **101**, 453—462.

MACLEOD, J. J. R., H. E. MAGEE, and C. B. PURVES, 1930: Selective absorption of carbohydrates. J. Physiol. (Brit.) **70**, 404—416.

MAGEE, H. E., and E. RIED, 1931: Absorption of glucose from alimentary canal. J. Physiol. (Brit.) **73**, 163—183.

MAIZELS, M., 1935: The permeation of erythrocytes by cations. Biochem. J. **29**, 1970—1982.

— 1948: Control of cations in erythrocytes. J. Physiol. (Brit.) **107**, 9 P—10 P.

— 1949: Cation control in human erythrocytes. J. Physiol. (Brit.) **108**, 247—263.

— 1951: Factors in active transport of cations. J. Physiol. (Brit.) **112**, 59—83.

— 1954: Cation transport in chicken erythrocytes. J. Physiol. (Brit.) **125**, 263—277.

MALM, M., 1940: Quantitative Bestimmungen der Permeabilität der Hefezellen für Fluor. Die Naturwissenschaften **28**, 723—724.

— 1948: Über die Permeabilität der Hefezellen und die von den permeierenden Stoffen, insbesondere Fluorwasserstoff, bedingten Plasmaveränderungen. Ark. Kemi Mineral. Geol. **25**, 1—187.

MARSH, J. B., and D. L. DRABKIN, 1947: Kidney phosphatase in alimentary hyperglycemia and phlorhizin glycosuria. A dynamic mechanism for renal threshold for glucose. J. biol. Chem. (Am.) **168**, 61—73.

MATHIEU, Fr., 1935: Die Resorption von Hexose-di- und -monophosphorsäure im Vergleich zu anderen Hexosen. Biochem. Z. **276**, 49—54.

MATTHEWS, D. M., and D. H. SMYTH, 1952: Stereochemically specific absorption of alanine from the intestine into the blood stream. J. Physiol. (Brit.) **116**, 20 P—21 P.

MELDAHL, K. F., und S. L. ØRSKOV, 1940: Photoelektrische Methode zur Bestimmung der Permeierungsgeschwindigkeit von Anelektrolyten durch die Membran von roten Blutkörperchen. Untersuchungen über die Gültigkeit des Fickschen Gesetzes für die Permeierungsgeschwindigkeit. Skand. Arch. Physiol. **83**, 266—280.

MEYER, D. K., 1951: Sodium flux through the gills of goldfish. Amer. J. Physiol. **165**, 580—587.

8*

Minibeck, H., 1939: Die selektive Zuckerresorption beim Kaltblüter und ihre Be-
    einflussung durch Nebennieren- und Hypophysenexstirpation. Arch. ges. Physiol.
    242, 344—353.
Monroy Oddo, A., and M. Esposito, 1951: Changes in the potassium content of
    sea urchin eggs on fertilization. J. gen. Physiol. (Am.) 34, 285—293.
Montgomery, H., and J. A. Pierce, 1937: The site of acidification of the urine
    within the renal tubule in amphibia. Amer. J. Physiol. 118, 144—152.
Mortensen, R. A., and K. E. Kellogg, 1944: The uptake of lead by blood cells
    as measured with a radioactive isotope. J. cellul. a. comp. Physiol. (Am.) 23,
    11—20.
Mudge, G. H., 1951 a: Studies on potassium accumulation by rabbit kidney slices:
    effect of metabolic activity. Amer. J. Physiol. 165, 113—127.
— 1951 b: Electrolyte and water metabolism of rabbit kidney slices: effect of
    metabolic inhibitors. Amer. J. Physiol. 167, 206—223.
— 1953: Electrolyte metabolism of rabbit-kidney slices: studies with radioactive
    potassium and sodium. Amer. J. Physiol. 173, 511—522.
— 1954: Renal Mechanisms of Electrolyte Transport, in Clarke. H. T., and
    D. Nachmansohn: Ion Transport across Membranes, New York.
— J. Foulks, and A. Gilman, 1948: The renal excretion of potassium. Proc. Soc.
    exper. Biol. a. Med. (Am.) 67, 545—547.
— — — 1949: Effect of urea diuresis on renal excretion of electrolytes. Amer. J.
    Physiol. 158, 218—230.
— — — 1950: Renal secretion of potassium during cellular dehydration. Amer. J.
    Physiol. 161, 159—166.
— and J. V. Taggart, 1950 a: Effect of 2, 4-dinitrophenol on renal transport
    mechanisms in the dog. Amer. J. Physiol. 161, 173—180.
— — 1950 b: Effect of acetate on the renal excretion of p-aminohippurate in the
    dog. Amer. J. Physiol. 161, 191—197.
Mueller, C. B., and A. B. Hastings, 1951: The rate of transfer of phosphorus
    across the red blood cell membrane. J. biol. Chem. (Am.) 189, 869—879.
Mullins, L. J., 1942: The permeability of yeast cells to radiophosphate. Biol. Bull.
    (Am.) 83, 326—333.
Nagano, J., 1902: Zur Kenntnis der Resorption einfacher, im besonderen sterio-
    isomerer Zucker im Dünndarm. Arch. ges. Physiol. 90, 389—404.
Nagel, H., 1934: Die Aufgabe der Excretionsorgane und der Kiemen bei der
    Osmoregulation von Carcinus maenas. Z. vergl. Physiol. 21, 468—491.
Nakazawa, F., 1922: Influence of phlorhizin on intestinal absorption. Tohoku J.
    exper. Med. 3, 288—294.
Nickerson, W. J., 1948: Riboflavin enhancement of radioactive phosphate exchange
    by yeasts. J. gen. Microbiol. 2, l. c.
— 1949: Dependence, in yeasts, of phosphate uptake and polymerization upon the
    occurrence of glucose polymerization. Experientia 5, 202—203.
— and L. J. Mullins, 1948: Riboflavin enhancement of radioactive phosphate
    exchange by yeasts. Nature 161, 939—940.
— and K. Zerahn. 1949: Accumulation of radioactive cobalt by dividing yeast
    cells. Biochim. Biophys. Acta 3, 476—483.
Öhnell. R.. and R. Höber, 1939: Effect of various poisons on absorption of
    sugars and some other non-electrolytes from normal and isolated artificially
    perfused intestine. J. cellul. a. comp. Physiol. (Am.) 13, 161—174.
Ørskov, S. L., 1935: Untersuchungen über den Einfluß von Kohlensäure und Blei
    auf die Permeabilität der Blutkörperchen für Kalium und Rubidium. Biochem.
    Z. 279, 250—261.
— 1945: Investigations on the permeability of yeast cells. Acta Path. Microbiol.
    Scand. 22, 523—559.
— 1948: Experiments on active and passive permeability of Bacillus coli communis.
    Acta Path. Microbiol. Scand. 25, 277—283.
— 1950: Experiments with substances which make bakers yeast absorb potassium.
    Acta Physiol. Scand. 20, 62—78.
Opie, E. L., 1949: The movement of water in tissues removed from the body and
    its relation to movement of water during life. J. exper. Med. (Am.) 89, 185—208.
— 1950: The effect of injury by toxic agents upon osmotic pressure maintained
    by cells of liver and of kidney. J. exper. Med. (Am.) 91, 285—294.
Paine, T. F. Jr., 1951: The similarity in action of bacitracin and penicillin on the
    staphylococcus. J. Bacter. (Am.) 61, 259—260.

PARK, C. R., 1952: in W. D. McELROY and B. GLASS: Phosphorus Metabolism. Vol. II, Baltimore.
— 1954: An effect of insulin on glucose metabolism by muscle. Fed. Proc. 13, 108.
— D. H. BROWN, M. CORNBLATH, W. H. DAUGHADAY, and M. E. KRAHL, 1952: The effect of growth hormone on glucose uptake by the isolated rat diaphragm. J. biol. Chem. (Am.) 197, 151—166.
— and W. H. DAUGHADAY, 1949: Effect of growth hormone on the glucose uptake and glycogen synthesis by the rat diaphragm. Fed. Proc. 9, 212—213.
— and L. H. JOHNSON, 1953: The effect of insulin on the distribution of free glucose in muscle. XIXth Internat. Physiol. Congr. 661.
PARPART, A. K., E. S. G. BARRON, and T. DEY 1947: Are -SH groups involved in the penetration of glycerol into human red cells? Biol. Bull. (Am.) 93, 199.
— and J. F. HOFFMAN, 1952: Acidity vs. acetylcholine and cation permeability of red cells. Fed. Proc. 11, 117.
— — 1954: Ion Permeability of the Red Cell, in CLARKE, H. T., and D. NACHMANSOHN, Ion Transport across Membranes, New York.
PERTZOFF, V., and C. L. GEMMILL, 1949: The effect of anesthetics on the uptake of radioactive phosphorus by human erythrocytes. J. Pharmacol. (Am.) 95, 106—115.
PITTS, R., 1943 a: A renal reabsorptive mechanism in the dog common to glycin and creatine. Amer. J. Physiol. 140, 156—167.
— 1943 b: A comparison of the renal reabsorptive processes for several amino acids. Amer. J. Physiol. 140, 535—547.
— and R. S. ALEXANDER, 1944: The renal reabsorptive mechanism for inorganic phosphate in normal and acidotic dogs. Amer. J. Physiol. 142, 648—662.
— J. L. AYER, and W. A. SCHIESS, 1948: The reabsorption and excretion of bicarbonate in normal man. Fed. Proc. 7, 94.
— W. D. LOTSPEICH, W. A. SCHIESS, and J. L. AYER, 1948: The renal regulation of acid-base balance in man. I. The nature of the mechanism for acidifying the urine. J. clin. Invest. (Am.) 27, 48—56.
PONDER, E., 1949: The rate of loss of potassium from human red cells in systems to which lysins have not been added. J. gen. Physiol. (Am.) 32, 461—479.
— 1950: Accumulation of potassium by human red cells. J. gen. Physiol. (Am.) 33, 745—757.
— 1951: Anomalous features of the loss of K from human red cells: results of extended observations. J. gen. Physiol. (Am.) 34, 359—372.
— and G. SASLOW, 1931: Measurement of red cell volume: alterations of cell volume in extremely hypotonic solutions. J. Physiol. (Brit.) 73, 267—296.
POPJÁK, G., 1950: Mechanism of absorption of inorganic phosphate from blood by tissue cells. Nature 166, 184—185.
POULSSON, L. T., 1930: On the mechanism of sugar elimination in phlorrhizin glycosuria. A contribution to the filtration-reabsorption theory on kidney function. J. Physiol. (Brit.) 69, 411—422.
PRANKERD, T. A. J., and K. I. ALTMAN, 1954: Phosphate partition and turnover in human red cells. Nature 173, 870—871.
PRESCOTT, D. M., and E. ZEUTHEN, 1953: Comparison of water diffusion and water filtration across cell surfaces. Acta Physiol. Scand. 28, 77—94.
PRICE, C. A., and R. E. DAVIES, 1954: Active transport of water by mitochondria. Biochem. J. 58, xvii.
PULVER, R., und F. VERZÁR, 1940 a: Der Zusammenhang von Kalium- und Kohlehydratstoffwechsel bei der Hefe. Helvet. Chim. Acta 23, 1087—1100.
— — 1940 b: Connexion between carbohydrate and potassium metabolism in the yeast cell. Nature 145, 823—824.
— — 1941: Kalium- und Kohlehydratstoffwechsel der Leukocyten. Helvet. Chim. Acta 24, 272—277.
RAKER, J. W., I. M. TAYLOR, J. M. WELLER, and A. B. HASTINGS, 1950: Rate of potassium exchange of the human erythrocyte. J. gen. Physiol. (Am.) 33, 691—702.
RAMSAY, J. A., 1951: Osmotic regulation in mosquito larvae: the role of the Malpighian tubules. J. exper. Biol. 28, 62—73.
— 1953: Active transport of potassium by the Malpighian tubules of insects. J. exper. Biol. 30, 358—369.
REISER, R., 1942: The lipids of the duodenal mucosa of swine during the absorption of fat. J. biol. Chem. (Am.) 143, 109—114.

Rice, L., J. Frieden, and M. Smith, 1953: Tubular action of mercurial diuretics. Amer. J. Physiol. **175**, 47—50.

Richards, A. G., and O. H. Schmitt, 1953: Asymmetrical penetration through the isolated cuticles of fly larvae. XIXth Internat. Physiol. Congr. 699—700.

Riggs, T. R., H. N. Christensen, and I. M. Palatine, 1952: Concentrating activity of reticulocytes for glycine. J. biol. Chem. (Am.) **194**, 53—55.

— B. Coyne, and H. N. Christensen, 1953: Intensification of the cellular accumulation of amino acids by pyridoxal. Biochim. Biophys. Acta **11**, 303—304.

— — — 1954: Amino acid concentration by a free cell neoplasm. Structural influences. J. biol. Chem. (Am.) **209**, 395—411.

Roberts, R. B., and I. Z. Roberts, 1950: Potassium metabolism of *Escherichia coli*. III. Interrelationship of potassium and phosphorus metabolism. J. cellul. a. comp. Physiol. (Am.) **36**, 15—39.

— — and D. B. Cowie, 1949: Potassium metabolism in *Escherichia coli*. II. Metabolism in the presence of carbohydrates and their metabolic derivatives. J. cellul. a. comp. Physiol. (Am.) **34**, 259—291.

Robinson, J. R., 1950 a: Osmoregulation in surviving slices of the kidneys of adult rats. Proc. roy. Soc., Lond. **B 137**, 378—402.

— 1950 b: Effect of 2, 4-dinitrophenol on osmoregulation in isolated kidney slices. Nature **166**, 989—990.

— 1952 a: Osmoregulation in surviving slices from the livers of adult rats (with a note on cloudy swelling). Proc. roy. Soc., Lond. **B 140**, 135—144.

— 1952 b: Total concentration of fixed base in cells of the renal cortex of the rat. Nature **169**, 713—714.

— 1953: The active transport of water in living systems. Biol. Rev. **28**, 158—194.

Ronkin, R. R., 1950 a: The uptake of radioactive phosphate by the excised gill of the mussel, *Mytilus edulis*. J. cellul. a. comp. Physiol. (Am.) **35**, 241—260.

— 1950 b: Effect of inhibitors on phosphate uptake in excised gills of the mussel (*Mytilus edulis*). Proc. Soc. exper. Biol. a. Med. (Am.) **73**, 41—44.

Rosenberg, T., 1948: On accumulation and active transport in biological systems. I. Thermodynamic considerations. Acta Chem. Scand. **2**, 14—33.

— and W. Wilbrandt, 1952: Enzymatic processes in cell membrane penetration. Internat. Rev. Cytol. **1**, 65—92.

Ross, E. J., 1952: The influence of insulin on the permeability of the blood-aqueous barrier to glucose. J. Physiol. (Brit.) **116**, 414—423.

— 1953: Insulin and the permeability of cell membranes to glucose. Nature **171**, 125.

Rothenberg, M. A., 1950: Studies on permeability in relation to nerve function. II. Ionic movements across axonal membranes, Biochim. Biophys. Acta **4**, 96—114.

Rothstein, A., 1954: Enzyme systems of the cell-surface involved in the uptake of sugars by yeast. Symp. Soc. exper. Biol. **8** (in press).

— and C. Demis, 1953: The relationship of the cell surface to metabolism. The stimulation of fermentation by extracellular potassium. Arch. Biochem. Biophys. **44**, 18—29.

— and L. H. Enns, 1946: The relationship of potassium to carbohydrate metabolism in baker's yeast. J. cellul. a. comp. Physiol. (Am.) **28**, 231—252.

— A. Frenkel, and C. Larrabee, 1948: The relationship of the cell surface to metabolism. III. Certain characteristics of the uranium complex with cell surface groups of yeast. J. cellul. a. comp. Physiol. (Am.) **32**, 261—274.

— and C. Larrabee, 1948: The relationship of the cell surface to metabolism. II. The cell surface of yeast as the site of inhibition of glucose metabolism by uranium. J. cellul. a. comp. Physiol. (Am.) **32**, 247—259.

— and R. Meier, 1951: The relationship of the cell surface to metabolism. VI. The chemical nature of uranium-complexing groups of the cell surface. J. cellul. a. comp. Physiol. (Am.) **38**, 245—270.

— — 1954: Unpublished observations.

— — and L. Hurwitz, 1951: The relationship of the cell surface to metabolism. V. The role of uranium-complexing loci of yeast in metabolism. J. cellul. a. comp. Physiol. (Am.) **37**, 57—81.

— — and T. Scharff, 1953: The relationship of the cell surface to metabolism. IX. The digestion of phosphorylated compounds by enzymes located on the surface of the intestinal cell. University of Rochester Atomic Energy Project, Report UR-237.

RUNNSTRÖM, J., 1939: Permeabilität und Stoffwechsel bei Hefe. Arch. exper. Zell-forsch. 22. 614—619.

— und E. SPERBER, 1938: Zur Kenntnis der Beziehungen zwischen Permeabilität und Stoffwechsel der Hefezelle. Biochem. Z. 298, 340—367.

RUSSO, H. F., L. D. WRIGHT, and H. R. SKEGGS, 1947: Renal clearance of essential amino acids: threonine and phenylalanine. Proc. Soc. exper. Biol. a. Med. (Am.) 65, 215—217.

SABBATANI, L., 1901: Détermination du point de congélation des organes animaux. J. Physiol. Path. gén. 3. 939—950.

SACKS, J., 1944 a: Radioactive phosphorus studies on hexosemonophosphate meta-bolism in resting muscle. Amer. J. Physiol. 142, 145—151.

— 1944 b: Some factors influencing phosphate turnover in muscle. Amer. J. Physiol. 142. 621—626.

— 1945: The effect of insulin on phosphorus turnover in muscle. Amer. J. Physiol. 143, 157—162.

— 1948: Mechanism of phosphate transfer across cell membranes. Cold Spring Harbor Symp. Quant. Biol. 13. 180—184.

— 1951: Phosphate transport and turnover in the liver. Arch. Biochem. 30, 423—437.

— and C. H. ALTSHULER, 1942: Radioactive phosphorus studies on striated and cardiac muscle metabolism. Amer. J. Physiol. 137, 750—760.

SAWYER, W. H., 1951: Effect of posterior pituitary extract on permeability of frog skin to water. Amer. J. Physiol. 164, 44—48.

SCHLIEPER, C., 1933: Über die osmoregulatorische Funktion der Aalkiemen. Z. vergl. Physiol. 18. 682—695.

SCHMIDT, G., L. HECHT, and S. J. THANHAUSER, 1949: The effect of potassium ions on the absorption of orthophosphate and the formation of metaphosphate by bakers' yeast. J. biol. Chem. (Brit.) 178. 733—742.

SCHMIDT-NIELSEN, K., 1946: Investigations on the fat absorption in the intestine. Acta Physiol. Scand. 12, Suppl. 37.

SCHÖNHEYDER, F., 1934: Über die Permeabilität der roten Blutkörperchen für Malonamid. Skand. Arch. Physiol. 71, 39—60.

SCHOFFENIELS. E., 1951: L'absorption du radiophosphore par la branchie isolée de l'anodonte. Arch. internat. Physiol. 59. 245—247.

SCHOFIELD, F. A., and H. B. LEWIS, 1947: A comparative study of the metabolism of α-alanine, β-alanine. serine and isoserine. J. biol. Chem. (Am.) 168, 439—445.

SCHWARTZ, W. B., and W. M. WALLACE, 1951: Electrolyte equilibrium during mercurial diuresis. J. clin. Invest. (Am.) 30, 1089—1104.

SCHWERIN, P., S. P. BESSMAN, and H. WAELSCH, 1950: The uptake of glutamic acid and glutamine by brain and other tissues of the rat and mouse. J. biol. Chem. (Am.) 184, 37—44.

SHANES, A. M., 1951: Potassium movement in relation to nerve activity. J. gen. Physiol. (Am.) 34, 795—807.

SHANNON, J. A., 1938: Tubular reabsorption of xylose in normal dog. Amer. J. Physiol. 122, 775—781.

— and S. FISHER, 1938: Renal tubular reabsorption of glucose in normal dog. Amer. J. Physiol. 122, 765—774.

SHEPPARD, C. W., and W. R. MARTIN, 1950: Cation exchange between cells and plasma of mammalian blood. I. Methods and application to potassium exchange in human blood. J. gen. Physiol. (Am.) 33. 703—722.

— — and G. BEYL, 1951: Cation exchange between cells and plasma of mammalian blood. II. Sodium and potassium exchange in the sheep, dog, cow, and man and the effect of varying the plasma potassium concentration. J. gen. Physiol. (Am.) 34, 411—429.

SHIDEMAN, F. E., and R. M. RENE, 1951: Succinate oxidation and Krebs cycle as an energy source for renal tubular transport mechanisms. Amer. J. Physiol. 166, 104—112.

SIVILLA, S. V., 1953: Effect of hypertonic solutions on intestinal absorption of selective and non-selective sugars. XIXth Internat. Physiol. Congr. 761—762.

SMITH, H. W., 1910: The absorption and excretion of water and salts by marine teleosts. Amer. J. Physiol. 93, 480—505.

SOLLNER, K., S. DRAY, E. GRIM, and R. NEIHOF, 1954: Electrochemical Studies with Model Membranes, in H. T. CLARKE and D. NACHMANSOHN: Ion Transport across Membranes. New York.

SOLOMON, A. K., 1952: The permeability of the human erythrocyte to sodium and potassium. J. gen. Physiol. (Am.) 36. 57—110.

Solomon, A. K. and G. L. Gold, 1953: Potassium transport in human erythrocytes: evidence for a three compartment system (in press).

Soulairac, A., 1947: La régulation neuro-endocrinienne de l'absorption intestinale des glucides. Ann. d'Endocr. **8**, 377—393.

— P. Desclaux, et J. Teysseyre, 1949: Étude histochimique de la phosphatase alcaline rénale. La régulation endocrinienne de la réabsorption tubulaire du glucose. Ann. d'Endocr. **10**, 535—546.

Spanner, D. C., 1954: The active transport of water under temperature gradients. Symp. Soc. exper. Biol. **8** (in press).

Sperry, W. M., and F. C. Brand, 1939: Absorption of water by liver slices from "physiological" saline solutions. Proc. Soc. exper. Biol. a. Med. (Am.) **42**, 147—150.

Spiegelman. S., M. D. Kamen, and M. Sussman, 1948: Phosphate metabolism and the dissociation of anaerobic glycolysis from synthesis in the presence of sodium azide. Arch. Biochem. **18**, 409—436.

Stadie, W. C., 1953: Studies on the action of insulin *in vitro*. XIXth Internat. Physiol. Congr. 24—28.

— 1954: Current concepts of the action of insulin. Physiol. Rev. **34**, 52—100.

— N. Haugaard, A. G. Hills, and J. B. Marsh, 1949: Hormonal influences on the chemical combination of insulin with rat muscle (diaphragm). Amer. J. med. Sci. **218**, 275—280.

— — and J. B. Marsh, 1951 a: Combination of insulin with muscle of the hypophysectomized rat. J. biol. Chem. (Am.) **188**, 167—172.

— — — 1951 b: Combination of epinephrine and 2, 4-dinitrophenol with muscle of the normal rat. J. biol. Chem. (Am.) **188**, 173—178.

— — and A. G. Hills, 1949: The chemical combination of insulin with muscle (diaphragm) of normal rat. Amer. J. med. Sci. **218**, 265—274.

— — and M. Vaughan, 1952: Studies of insulin binding with isotopically labeled insulin. J. biol. Chem. (Am.) **199**, 729—739.

Stamler, J., 1951: Failure of tubular reabsorptive loads of ascorbic acid or amino acids to affect renal handling of sodium and potassium. Amer. J. Physiol. **165**, 109—112.

Stanbury, S. W., and G. H. Mudge, 1953: Potassium metabolism of liver mitochondria. Proc. Soc. exper. Biol. a. Med. (Am.) **82**, 675—681.

Steggerda. F. R., 1931: The relation of pitressin to water interchange in frogs. Amer. J. Physiol. **98**, 255—261.

Steinbach, H. B., 1940: Sodium and potassium in frog muscle. J. biol. Chem. (Am.) **133**, 695—701.

— 1951: Permeability. Ann. Rev. Physiol. **13**, 21—40.

— 1951: Sodium extrusion from isolated frog muscle. Amer. J. Physiol. **167**, 284—287.

— 1952: On the sodium and potassium balance of isolated frog muscles. Proc. Nat. Ac. Sci. **38**, 451—455.

Stern, J. R., L. V. Eggleston, R. Hems, and H. A. Krebs, 1949: Accumulation of glutamic acid in isolated brain tissue. Biochem. J. **44**, 410—418.

Stewart, D. R., and M. H. Jacobs, 1932 a: The effect of fertilization on the permeability of the eggs of *Arbacia* and *Asterias* to ethylene glycol. J. cellul. a. comp. Physiol. (Am.) **1**, 83—92.

— — 1932 b: The permeability of the egg of *Arbacia* to ethylene glycol at different temperatures. J. cellul. a. comp. Physiol. (Am.) **2**, 275—283.

Taggart, J. V., and R. P. Forster, 1950: Renal tubular transport: effect of 2, 4-dinitrophenol and related compounds on phenol red transport in the isolated tubules of the flounder. Amer. J. Physiol. **161**, 167—172.

Taylor. E. S., 1947: The assimilation of amino-acids by bacteria. 3. Concentration of free amino-acids in the internal environment of various bacteria and yeasts.

Taylor, I. M., and J. M. Weller, 1950: Studies on the permeability of human erythrocytes to potassium. Biol. Bull. (Am.) **99**, 311.

— J. M. Weller, and A. B. Hastings, 1952: Effect of cholinesterase and choline acetylase inhibitors on the potassium concentration gradient and potassium exchange of human erythrocytes. Amer. J. Physiol. **168**, 658—668.

Teorell, T., 1953: Transport processes and electrical phenomena in ionic membranes. Progress in Biophysics and Biophysical Chemistry **3**, 305—369.

Terner, C., L. V. Eggleston, and H. A. Krebs, 1950: The role of glutamic acid in the transport of potassium in brain and retina. Biochem. J. **47**, 139—149.

THOMPSON, V., and A. TICE, 1941: Action of drugs beneficial in myasthenia gravis. I. Effect of prostigmine and guanidine on serum and muscle potassium. J. Pharm. exper. Ther. **73**, 455—462.

TOSTESON, D. C., and E. T. DUNHAM, 1954: Effect of sickling on sodium and cesium transport. Fed. Proc. **13**, 523.

TRIMBLE, H. C., B. W. CAREY Jr., and S. J. MADDOCK, 1933: The rate of absorption of glucose from the gastrointestinal tract of the dog. J. biol. Chem. (Am.) **100**, 125—138.

USSING, H. H., 1943 a: The nature of the amino nitrogen of red corpuscles. Acta Physiol. Scand. **5**, 335—351.

— 1943 b: On the partition of certain amino acids between blood and tissues. Acta Physiol. Scand. **6**, 222—232.

— 1945: The reabsorption of glycine and other amino acids in the kidneys of man. Acta Physiol. Scand. **9**, 193—213.

— 1947: Interpretation of the exchange of radio-sodium in isolated muscle. Nature **160**, 262—263.

— 1948: The use of tracers in the study of active ion transport across animal membranes. Cold Spring Harbor Symp. Quant. Biol. **13**, 193—200.

— 1949: The active ion transport through the isolated frog skin in the light of tracer studies. Acta Physiol. Scand. **17**, 1—37.

— 1952: Some aspects of the application of tracers in permeability studies. Adv. in Enzymol. **13**, 21—65.

— 1953: Transport through biological membranes. Ann. Rev. Physiol. **15**, 1—20.

— 1954: Active transport of inorganic ions. Symp. Soc. exper. Biol. **8** (in press).

— and K. ZERAHN, 1951: Active transport of sodium as the source of electric current in the short-circuited isolated frog skin. Acta Physiol. Scand. **23**, 110—127.

VAN SLYKE, D. D., and G. M. MEYER, 1913: The fate of protein digestion products in the body. III. The absorption of amino-acids from the blood by the tissues. J. biol. Chem. (Am.) **16**, 197—212.

VERZÁR, F., 1935: Die Rolle von Diffusion und Schleimhautaktivität bei der Resorption von verschiedenen Zuckern aus dem Darm. Biochem. Z. **276**, 17—27.

— und L. LASZT, 1934 a: Untersuchungen über die Resorption von Fettsäuren. Biochem. Z. **270**, 24—34.

— — 1934 b: Hemmung der Fettresorption durch Monoiodessigsäure und Phlorrhizin. Biochem. Z. **270**, 35—43.

— — 1935 a: Die Hemmung der Fettresorption durch Phlorrhizin. Biochem. Z. **276**, 1—10.

— — 1935 b: Die Hemmung der Fettresorption nach Exstirpation der Nebennieren. Biochem. Z. **276**, 11—16.

— — 1935 c: Die Resorption aus dem Darm von isotonischen Lösungen von Glucose und Sorbose, verglichen mit der von Natriumsulfat. Biochem. Z. **276**, 28—39.

— and J. C. SOMOGYI, 1939: Connexion between carbohydrate and potassium metabolism in normal and adrenalectomized animale. Nature **144**, 1014—1015.

— — 1940: Liberation of potassium from muscle by acetylcholine and muscle contraction and its absence after adrenalectomy. Nature **145**, 781.

— and V. WENNER, 1948: The influence in vitro of deoxycorticosterone on glycogen formation in muscle. Biochem. J. **42**, 35—41.

— und H. WIRZ, 1937: Weitere Untersuchungen über die Bedingungen der selektiven Glucoseresorption. Biochem. Z. **292**, 174—181.

VILLEE, C. A., and A. B. HASTINGS, 1949: The metabolism of C^{14}-labelled glucose by the rat diaphragm in vitro. J. biol. Chem. (Am.) **179**, 673—687.

— M. LOWENS, M. GORDON, E. LEONARD, and A. RICH, 1949: The incorporation of P^{32} into the nucleoproteins and phosphoproteins of the developing sea urchin embryo. J. cellul. a. comp. Physiol. (Am.) **33**, 93—112.

VISSCHER, M. B., E. S. FETCHER Jr., C. W. CARR, H. P. GREGOR, M. S. BUSHEY, and D. E. BAKER, 1944: Isotopic tracer studies on the movement of water and ions between intestinal lumen and blood. Amer. J. Physiol. **142**, 550—575.

— and R. R. ROEPKE, 1945: Osmotic and electrolyte concentration relationships during absorption of salt solutions from ileal segments. Amer. J. Physiol. **144**, 468—476.

— R. H. VARCO, C. W. CARR, R. B. DEAN, and D. ERICKSON, 1944: Sodium ion movement between the intestinal lumen and the blood. Amer. J. Physiol. **141**, 488—505.

Walker, A. M., P. A. Bott, J. Oliver, and M. C. MacDowell, 1941: The collection and analysis of fluid from single nephrons of the mammalian kidney. Amer. J. Physiol. 134, 580—595.
— and C. L. Hudson, 1937 a: The reabsorption of glucose from the renal tubule in amphibia and the action of phlorizin upon it. Amer. J. Physiol. 118, 130—143.
— — 1937 b: The rôle of the tubule in the excretion of inorganic phosphates by the amphibian kidney. Amer. J. Physiol. 118, 167—173.
— — T. Findley Jr., and A. N. Richards, 1937: The total molecular concentration and the chloride concentration of fluid from different segments of the renal tubule of amphibia. Amer. J. Physiol. 118, 121—129.
Webb, D. A., 1940: Ionic regulation in Carcinus maenas. Proc. roy. Soc., Lond. B 129, 107—136.
Welt, L. G., J. Orloff, D. M. Kydd, and J. E. Oltman, 1950: An example of cellular hyperosmolarity. J. clin. Invest. (Am.) 29, 935—939.
Wertheimer, E., 1933: Phlorrhizinwirkung auf die Zuckerresorption. Arch. ges. Physiol. 233, 514—528.
— Über die ersten Anfänge der Zuckerassimilation. Versuche an Hefezellen. Protoplasma 21, 522—560.
Wesson, L. G. Jr., W. E. Cohn, and A. M. Brues, 1949: The effect of temperature on potassium equilibria in chick embryo muscle. J. gen. Physiol. (Am.) 32. 511—524.
West, C. D., S. A. Kaplan, S. J. Fomon, and S. Rapoport, 1952: Urine flow and solute excretion during osmotic diuresis in hydrated dogs: role of distal tubule in the production of hypotonic urine. Amer. J. Physiol. 170, 239—254.
Westenbrink, H. G. K., 1934: Über die Anpassung der Darmresorption an die Zusammensetzung der Nahrung. Arch. Néerl. Physiol. 19, 563—583.
— 1937: Relative velocities of the absorption of different sugars from the intestine of rat and pigeon. Nature 138, 203—204.
— und K. Gratama, 1937: Über die Spezifität der Resorption einiger Monosen aus dem Darme der Ratte und der Taube. Arch. Néerl. Physiol. 21, 433—454.
Whittam, R., and R. E. Davies, 1953 a: Transport of water, sodium, potassium, and α-ketoglutarate in kidney cortex slices. Biochem. J. 54, vii.
— — 1953 b: Measurements of the turnover-rates of sodium and potassium in kidney cortex slices. Biochem. J. 54, vii—viii.
— — 1953 c: Active transport of water, sodium, potassium and α-oxoglutarate by kidney-cortex slices. Biochem. J. 55, 880—888.
— — 1954: Relations between metabolism and the rate of turnover of sodium and potassium in guinea pig kidney-cortex slices. Biochem. J. 56, 445—453.
Wick, A. N., and D. R. Drury, 1951 a: Action of insulin on the permeability of cells to sorbitol. Amer. J. Physiol. 166, 421—423.
— — 1951 b: Does concentration of glucose in extracellular fluid influence its utilization by the tissues? Amer. J. Physiol. 167, 359—363.
— — 1953 a: Action of insulin on volume of distribution of galactose in the body. Amer. J. Physiol. 173, 229—232.
— — 1953 b: Influence of glucose concentration on the action of insulin. Amer. J. Physiol. 174, 445—447.
Widdas, W. F., 1952 a: Inability of diffusion to account for placental glucose transfer in the sheep. J. Physiol. (Brit.) 115, 36 P.
— 1952 b: Inability of diffusion to account for placental transfer in the sheep and consideration of the kinetics of a possible carrier transfer. J. Physiol. (Brit.) 118, 23—39.
— 1953 a: Kinetics of glucose transfer across the human erythrocyte membrane. J. Physiol. (Brit.) 120, 23 P—24 P.
— 1953 b: Hexose permeability of mammalian foetal erythrocytes. XIXth Internat. Physiol. Congr. 885—886.
— 1954: Facilitated transfer of hexoses across the human erythrocyte membrane. J. Physiol. (Brit.) 125, 163—180.
Wilbrandt, W., 1938: Die Permeabilität der roten Blutkörperchen für einfache Zucker. Arch. ges. Physiol. 241. 302—309.
— 1940 a: Die Abhängigkeit der Ionenpermeabilität der Erythrocyten vom glykolytischen Stoffwechsel. Arch. ges. Physiol. 243, 519—536.
— 1940 b: Die Ionenpermeabilität der Erythrocyten in Nichtleiterlösungen. Arch. ges. Physiol. 243, 537—556.

WILBRANDT, W., 1941: Die Wirkung von Schwermetallsalzen auf die Erythrocyten-permeabilität für Glyzerin. Arch. ges. Physiol. **244**, 637—643.
— 1947: Die Wirkung des Phlorrhizins auf die Permeabilität der menschlichen Erythrocyten für Glukose und Pentosen. Helvet. Physiol. Acta **5**, C 64—C 65.
— 1950: Permeabilitätsprobleme. Arch. exper. Path. Pharmakol. **212**, 9—29.
— E. GUENSBERG, und H. LAUENER, 1947: Der Glukoseeintritt durch die Erythro-cytenmembran. Helvet. Physiol. Pharmacol. Acta **5**, C 20—C 22.
— und L. LASZT, 1933: Untersuchungen über die Ursachen der selektiven Resorp-tion der Zucker aus dem Darm. Biochem. Z. **259**, 398—417.
— und T. ROSENBERG, 1950: Weitere Untersuchungen über die Glukosepenetration durch die Erythrocytenmembran. Helvet. Physiol. Pharmacol. Acta **8**, C 82—C 83.
— — 1951: Die Kinetik des enzymatischen Transports. Helvet. Physiol. Acta **9**, C 86—C 87.
WILMER, H. A., 1944: Renal phosphatase. The correlation between the functional activity of the renal tubule and its phosphatase content. Arch. Path. **37**, 227—237.
WILSON, R. H., 1932: The effect of phlorhizin on the rate of absorption from the gastrointestinal tract of the white rat. J. biol. Chem. (Am.) **97**, 497—502.
WILSON, T. H., 1954: Ionic permeability and osmotic swelling of cells. Science **120**, 104—105.
— and G. WISEMAN, 1954: The use of sacs of everted small intestine for the study of the transference of substances from the mucosal to the serosal surface. J. Physiol. (Brit.) **123**, 116—125.
WIRZ, H., B. HARGITAY und W. KUHN, 1951: Lokalisation des Konzentrierungs-prozesses in der Niere durch direkte Kryoscopie. Helvet. Physiol. Acta **9**, 196—207.
WISEMAN, G., 1951: Active stereochemically selective absorption of amino-acids from rat small intestine. J. Physiol. (Brit.) **114**, 7 P—8 P.
— 1953: Absorption of amino-acids using an *in vitro* technique. J. Physiol. (Brit.) **120**, 63—72.
WIX, G., I. BONTA, L. GYÖRGY, and G. FEKETE, 1952: Hormonal influences on glucose resorption from the intestines. V. Contributions to the mechanism of insulin effect. Acta Physiol. Ac. Sci. Hungar. **3**, 59—68.
— G. FEKETE, and I. HORVÁTH, 1951: Hormonal influences on glucose resorption from the intestines. III. The effect of adrenalin and the resorption of glucose. Acta Physiol. Ac. Sci. Hungar. **2**, 451—457.
WOOD, E. H., 1941: Glucose reabsorption in the amphibian kidney. Amer. J. Physiol. **133**, P 497.
— D. A. COLLINS, and G. K. MOE, 1940: Electrolyte and water exchanges between mammalian muscle and blood in relation to activity. Amer. J. Physiol. **128**, 635—652.
WRIGHT, L. D., H. F. RUSSO, H. R. SKEGGS, E. A. PATCH, and K. H. BEYER, 1947: The renal clearance of essential amino acids: arginine, histidine, lysine, and methionine. Amer. J. Physiol. **149**, 130—134.

Protoplasmatologia Handbuch der Protoplasmaforschung

Unter Mitwirkung hervorragender internationaler Fachleute

herausgegeben von **L. V. Heilbrunn**, Philadelphia, und **F. Weber**, Graz

In 14 Bänden

Das Handbuch erscheint in selbständigen Einzelveröffentlichungen, die zu Bänden vereinigt werden. Jeder selbständig erscheinende Handbuchteil ist einzeln käuflich. Bei Verpflichtung zur Abnahme des gesamten Handbuches, bei Vorbestellung der einzelnen Teile sowie für Abonnenten der Zeitschrift „Protoplasma" ermäßigt sich der Preis um 20%

Bisher sind erschienen:

Die makromolekulare Chemie und ihre Bedeutung für die Protoplasmaforschung. Von Prof. Dr. phil., Dr.-Ing. e. h., Dr. rer. nat. h. c., Dr. (C) h. c. **Hermann Staudinger**, und Dr. phil., Mag. rer. nat. **Magda Staudinger**, beide Staatliches Forschungsinstitut für makromolekulare Chemie der Universität Freiburg i. Br. **Band I. Grundlagen. 1.** Die makromolekulare Chemie und ihre Bedeutung für die Protoplasmaforschung. Mit 27 Textabbildungen. IV, 73 Seiten. Gr.-8°. 1954.
S 117.—, DM 19.50, sfr. 20.—, $ 4.65

Die submikroskopische Struktur des Cytoplasmas. Von Prof. Dr. **A. Frey-Wyssling**, Institut für Allgemeine Botanik der Eidg. Technischen Hochschule, Zürich. **Band II. Cytoplasma. A.** Morphologie. 2. Die submikroskopische Struktur des Cytoplasmas. Mit 90 Textabbildungen. IV, 244 Seiten. Gr.-8°. 1955.
S 255.—, DM 42.50, sfr. 43.50, $ 10.10

The pH of Plant Cells. By Prof. Dr. **James Small**, The Queen's University Belfast, Department of Botany. With 3 figures. 116 pages. — **The pH of Animal Cells.** By Prof. Dr. **Floyd J. Wiercinski**, Hahnemann Medical College, Department of Physiology, Philadelphia. With 7 figures. 56 pages. Gr.-8°. 1955. **Band II. Cytoplasma. B.** Chemie. 2. Cytochemie und Histochemie. c. The pH of Plant Cells. The pH of Animal Cells.
S 270.—, DM 45.—, sfr. 46.—, $ 10.70

The Enzymology of the Cell Surface. By **Aser Rothstein**, Rochester, New York. With 21 figures. 86 pages. — **Tension at the Cell Surface.** By **E. Newton Harvey**, Princeton, New Jersey. With 13 figures. 30 pages. Gr.-8°. 1954. **Band II. Cytoplasma. E.** Cytoplasma-Oberfläche. 4. The Enzymology of the Cell Surface. 5. Tension at the Cell Surface.
S 168.—, DM 28.—, sfr. 28.80, $ 6.70

Chemistry and Physiology of Mitochondria and Microsomes. By **Olov Lindberg**, Ph. D., and **Lars Ernster**, Ph. D., beide Wenner-Gren's Institute, Stockholm. **Band III. Cytoplasma-Organellen. A.** Chondriosomen, Mikrosomen, Sphaerosomen. 4. Chemistry and Physiology of Mitochondria and Microsomes. With 32 figures. IV, 136 pages. Gr.-8°. 1954.
S 204.—, DM 34.—, sfr. 34.80, $ 8.10

Endomitose und endomitotische Polyploidisierung. Von Prof. Dr. **Lothar Geitler**, Botanisches Institut der Universität Wien. **Band VI. Kern- und Zellteilung. C.** Endomitose und endomitotische Polyploidisierung. Mit 44 Textabbildungen. IV, 89 Seiten. Gr.-8°. 1953.
S 140.—, DM 23.50, sfr. 24.10, $ 5.60

Red Cell Structure and its Breakdown. By Prof. Dr. **Eric Ponder**, The Nassau Hospital, Mineola, N. Y. **Band X. Pathologie des Protoplasmas. 1.** Red Cell Structure and Its Breakdown. With 58 figures. IV, 124 pages. Gr.-8°. 1955.
S 240.—, DM 40.—, sfr. 40.90, $ 9.50

Protoplasmatische Pflanzenanatomie. Von Dr. **Lotte Reuter**, Privatdozent an der Universität Wien. **Band XI. Vergleichende Protoplasmatik. 2.** Protoplasmatische Pflanzenanatomie. Mit 64 Textabbildungen. IV, 131 Seiten. Gr.-8°. 1955.
S 204.—, DM 34.—, sfr. 34.80, $ 8.10